T0205922

Beginning Azure Cognitive Services

Data-Driven Decision Making Through Artificial Intelligence

Alicia Moniz
Matt Gordon
Ida Bergum
Mia Chang
Ginger Grant

Apress®

Beginning Azure Cognitive Services: Data-Driven Decision Making Through Artificial Intelligence

Alicia Moniz
Houston, TX, USA

Ida Bergum
Oslo, Norway

Ginger Grant
Scottsdale, AZ, USA

Matt Gordon
Lexington, KY, USA

Mia Chang
Berlin, Berlin, Germany

ISBN-13 (pbk): 978-1-4842-7175-9
https://doi.org/10.1007/978-1-4842-7176-6

ISBN-13 (electronic): 978-1-4842-7176-6

Managing Director, Apress Media LLC: Welmoed Spahr
Acquisitions Editor: Jonathan Gennick
Development Editor: Laura Berendson
Coordinating Editor: Jill Balzano

Cover image designed by Freepik (www.freepik.com)

Distributed to the book trade worldwide by Springer Science+Business Media LLC, 1 New York Plaza, Suite 4600, New York, NY 10004. Phone 1-800-SPRINGER, fax (201) 348-4505, e-mail orders-ny@springer-sbm.com, or visit www.springeronline.com. Apress Media, LLC is a California LLC and the sole member (owner) is Springer Science + Business Media Finance Inc (SSBM Finance Inc). SSBM Finance Inc is a Delaware corporation.

For information on translations, please e-mail booktranslations@springernature.com; for reprint, paperback, or audio rights, please e-mail bookpermissions@springernature.com.

Apress titles may be purchased in bulk for academic, corporate, or promotional use. eBook versions and licenses are also available for most titles. For more information, reference our Print and eBook Bulk Sales web page at http://www.apress.com/bulk-sales.

Any source code or other supplementary material referenced by the author in this book is available to readers on GitHub via the book's product page, located at www.apress.com/9781484271759. For more detailed information, please visit http://www.apress.com/source-code.

Printed on acid-free paper

Thanks to every person who has attended my presentations around the world. Thanks especially to those that asked questions or provided feedback that prompted me to work harder to find an answer.

—Matt Gordon

To the MVP program and the community, where our energy comes from.

—Mia Chang

Table of Contents

About the Authors

Alicia Moniz is a Microsoft AI MVP, #KafkaOnAzure Evangelista, and an active supporter of women in technology. She authors the blog HybridDataLakes.com, a blog focused on cloud data learning resources, and produces content for the YouTube channel #KafkaOnAzure. She is also on the Leadership Board for the Global AI community and is an organizer for the Global AI Bootcamp – Houston Edition, a Microsoft AI–sponsored event. Alicia is active in the Microsoft User Group community and enjoys speaking on AI, SQL Server, #KafkaOnAzure, and personal branding for women in technology topics.

Matt Gordon is a Microsoft Data Platform MVP and has worked with SQL Server since 2000. He is the leader of the Lexington, KY, Data Technology Group and a frequent domestic and international community speaker. He's an IDERA ACE alumnus and 2021 Friend of Redgate. His original data professional role was in database development, which quickly evolved into query tuning work that further evolved into being a DBA in the healthcare realm. He has supported several critical systems utilizing SQL Server and managed dozens of 24/7/365 SQL Server implementations. Following several years as a consultant, he is now the Director of Data and Infrastructure for rev.io, where he is implementing data governance, DevOps, and performance improvements enterprise-wide.

Ida Bergum is a Microsoft Data Platform MVP and is currently working for Avanade Norway as a solution architect for the Data & AI Market Unit as well as leading Avanade's Community of Practice for Power BI globally. Ida has been advising on, architecting, and implementing modern data platform and BI solutions on Azure since she started at Avanade in 2015. She aims at advising and delivering great analytics experiences and data-smart solutions for clients while sharing those experiences with the community. Her spare time is spent contributing to Data Platform User Group Norway among others, posting to her Twitter account, and, whenever she gets the opportunity, cross-country skiing.

Mia Chang works as a solution architect specializing in Machine Learning (ML) for a tech company in Berlin, Germany. She coordinates with customers on their development projects and delivers architecture leveraging services in the cloud. Her Machine Learning journey started with her studies in the fields of applied mathematics and computer science focused on algorithm research of strategic board games. During her 7 years of professional journey, she has worked as a data scientist and become proficient in the ML project lifecycle. Apart from this, she is passionate about working on open source projects and has received the Microsoft AI MVP Award from 2017 to 2020. Her first book *Microsoft AI MVP Book* was published in 2019. She also dedicates her time to organizing communities and speaks at meetups and conferences. In addition to tech-related hobbies, she loves gardening, hiking, and traveling. Visit her blog at https://bymiachang.com/.

Ginger Grant is a consultant who shares what she has learned while working with data technology with clients by providing training to people around the world. As a Microsoft MVP in Data Platform and a Microsoft Certified Trainer, she is proficient in creating solutions using Power BI and the Microsoft Azure Data Stack components including Databricks, Data Factory, Data Lakes, Data Analytics, Synapse, and Machine Learning. Ginger co-authored *Exam Ref 70-774 Perform Cloud Data Science with Azure Machine Learning* and has a number of current exam certifications. When not working, she maintains her blog and spends time on Twitter.

About the Technical Reviewer

Jamie Maguire is passionate about using AI technologies to help advance systems in a wide range of organizations.

He has collaborated on many projects including working with Twitter, National Geographic, and the University of Michigan. Jamie is a keen contributor to the technology community and has gained global recognition for articles he has written and software he has built.

He is the founder of a Twitter analytics and productivity platform called Social Opinion, STEM Ambassador, and Code Club volunteer, inspiring interest at the grassroots level. Jamie shares his story and expertise at speaking events, on social media, and through podcasts and interviews.

He has co-authored a book with 16 fellow MVPs demonstrating how Microsoft AI can be used in the real world and regularly publishes material to encourage and promote the use of AI and .NET technologies.

Find out more about Jamie at www.jamiemaguire.net or Social Opinion here: www.socialopinion.co.uk.

Acknowledgments

Thank you to Laurie Carr for finding me at my first SQL PASS conference and encouraging me to get involved. My life's path was altered the day your kindness found me. I will forever be grateful to my #SQLfamily and my family at Avanade for the opportunities that I have been given to grow and contribute to the Microsoft community. Thank you, mom, for instilling in me my capabilities for hard work and lifelong learning. And my eternal gratitude to my family – Tiana, Isaiah, and David – for their patience in giving me the space to chase my passions and, of course, write this book!

—Alicia Moniz

I'd like to thank Roger Bennett, Michael Davies, and Jonathan Williamson of *Men in Blazers*, whose "crap" podcast inspired me to look into using sentiment analysis in Cognitive Services for the Premier League Mood Table on the show. Thanks also go to the best boss I've ever had, Brad Ball, for his sentiment analysis blog that started me on the journey that led to this book. Finally, thanks to my wife Jennifer for putting up with me while this book came together, my daughter Anna for complaining that the donut bot doesn't bring donuts, and my son Eric for noting that AI is "really just a bunch of if-then statements that run quickly."

—Matt Gordon

For my mom and dad, #Data #PowerPlatform community friends, and Avanade for inspiring and supporting me.

—Ida Bergum

To my grandma and my family, for your unconditional love. To my co-authors, for being part of this journey together.

—Mia Chang

ACKNOWLEDGMENTS

I would like to thank a number of people who made this book possible, including my fellow authors and Apress. I would also like to thank Chris Johnson for getting her started with computers, my dad for encouraging me to study computers in college and I hope this book will inspire my two favorite students, Stuart and Corinne, to be inspired to see their names on a cover of a book someday.

—Ginger Grant

Introduction

During the holiday season of 2019, Microsoft debuted an ad campaign featuring a little girl using her tablet to have a conversation with reindeer in her yard. The tablet was translating reindeer grunts into English the little girl could understand. The graphic at the end said "Microsoft AI," but what the commercial was really showing you was some of the ability of Cognitive Services (although Reindeer is not a currently supported language – must still be in preview!).

Unfortunately, we will not be presenting a demo within this book of how to use a device to speak to reindeer as they're unable to sign a book contract. The good news is that we will be presenting to you a wide variety of demos, labs, and case studies about what you can do with Cognitive Services. We five authors came together as a group of AI and data professionals that have been fortunate enough to speak all around the world about the exciting things that can be done with this suite whether or not you're in data or AI or somewhere between the two. We all followed different paths to our work in AI and our work on this book – some as data professionals and some as AI-focused professionals.

What all of us hope you take away from this book is that no matter your career path or role, the abilities of Cognitive Services are within your reach. This book will walk you through a variety of no-code, low-code, and code-heavy methods of using these Cognitive Services to put the power of AI in your apps, websites, personal projects, professional projects, and everything in between. All of us have had people help us along the way as we learned these technologies, and our hope is that this book provides a useful kick start to you on your AI journey whether you're embarking on that for fun, for money, or both.

Intended Audience

The intended audience for this book is really anybody who is interested in playing around with AI technologies. Whether you're interested in applying AI to language, vision, speech, chatbots, or anything surrounding those abilities, this book is for you.

Many of the examples contained in this book have been refined over the years as we've presented and discussed these at various conferences, events, and workshops. If you are looking for a fun side project, this book should help you. If you're looking to bolster your understanding of how some of these technologies work, this book is for you. If you are looking to add some AI projects to your resume, this book will help you do that as well.

Finally, we would be remiss if we did not also mention the ethical issues in play with how these algorithms are created, trained, and tested. Our book covers that as well because it's truly important, especially if you plan on using this book as a jump start to a professional AI journey.

AI has been, is, and will be a part of our world for a very long time. If you're interested in how it works and interested in being a part of seeing what it can do for you and others, this book is for you. Let's get started!

CHAPTER 1

Introducing Cognitive Services

Azure Cognitive Services is a suite of services offered on Microsoft Azure that allows developers to integrate and embed AI capabilities within their applications. Microsoft Azure, for those new to cloud computing, is the overall brand for Microsoft's cloud computing offerings. Azure's AI capabilities allow users to see, hear, speak, search, understand, and accelerate decision making. Using Cognitive Services, you can detect emotion in customers' faces using the Vision API, instantly analyze customer communication to quickly react to impending issues using Text Analytics within the Language API, and combine the Language and Speech APIs to communicate with a customer in their native language – and that's just a taste of what can be done with Cognitive Services.

Cognitive Services APIs provide access to pre-trained machine learning models and require very little coding to integrate and operationalize. For the developer/data guru/ architect just getting started with Azure Cognitive Services, this chapter will provide an overview of functionality available within the Azure Cognitive Services suite. Throughout this book, real-world examples and common design patterns used when architecting AI-enhanced solutions will be covered to solidify understanding of how services integrate with existing systems.

Cognitive Services APIs are for the masses and are not the solution for a data scientist who wants to build their own machine learning model from scratch. Rather, the APIs are a commoditized form of Machine Learning that abstracts the storage-level coding and work that goes into building complicated machine learning models. Additionally, because the services are hosted on Microsoft compute resources and exposed via Application Programming Interfaces (APIs), results are accessible to both programmers and data practitioners with very few lines of code. Moreover, interaction with the APIs can be performed via code or the web interface provided by Microsoft. Programmers

© Alicia Moniz, Matt Gordon, Ida Bergum, Mia Chang, Ginger Grant 2021
A. Moniz et al., *Beginning Azure Cognitive Services*, https://doi.org/10.1007/978-1-4842-7176-6_1

with little programming background and little exposure to Machine Learning can both build and interact with a Cognitive Services model within a day! This chapter will largely be an overview of the capabilities of and resources for the Azure Cognitive Services suite, but let us kick it off with a few examples of how the power of Azure Cognitive Services can be applied to some interesting issues and problems.

Understanding Cognitive Services

Artificial Intelligence (AI) is certainly a buzzword these days, and it seems like we are surrounded by products and services claiming to be "powered by AI" or AI driven. While some of that is true, much of it is merely a sales pitch. Artificial Intelligence makes everything sound smarter and more capable, making it part of an effective sales pitch, but because of the complex computer science involved in its implementation, the average person is not going to closely examine whether or not the product or service is actually using AI as opposed to "a computer being involved somewhere along the way." Real AI is not as ubiquitous as marketing communications would have you think – mostly because until recently implementing AI has been very complicated and required highly technical people.

Azure Cognitive Services puts the power of AI directly into our hands as technical professionals or hobbyists. That sounds cool, but what can we actually do with this newfound power? First, we must understand what we are working with – the section titled "What Is a Cognitive Service?" gives a brief technical overview of what the guts of a Cognitive Service look like. Following that, let us review a few real-world scenarios and examples where the power of Azure Cognitive Services helped respond to a situation.

What Is a Cognitive Service?

A Cognitive Service is not something you can summon with a spell or incantation. There are two ways to invoke a Cognitive Service: you can call it using REST API calls, or you can interact with it via language-based Software Development Kits (SDKs). These language-based SDKs are currently available in C#, Go, Java, JavaScript, Python, and R. In so many words, that means that you need to have basic programming language knowledge to use the services. This book will include several examples of interacting with Azure Cognitive Services using a variety of popular programming languages. What are we invoking, though? At its most basic level, a Cognitive Service provides a trained machine learning model for you. A REST API call or SDK interaction brings data into this picture and possibly a choice of ML algorithm as well (depending on the Cognitive Service itself). These services can be invoked in minutes using data and basic API calls.

What types of questions can a Cognitive Service answer, or what sort of problems can it solve? Throughout this book, we will provide a number of examples, but, after a listing of the contents of the Azure Cognitive Services suite itself, let us reference a few interesting examples to kick-start your journey of learning how to implement Azure Cognitive Services and expand your professional skillset.

Reference Architecture

The Cognitive Services are APIs that are hosted on Microsoft-managed Azure computing resources. The services provide endpoints to machine learning models that Microsoft has curated training datasets for, trained and validated to have an expected outcome as far as accuracy and precision. Before the advent of Azure Cognitive Services, you would have needed to employ experienced machine learning experts to create, curate, and train these models and then provide programmatic access to those models. Microsoft has done all of this for us, which gives us the ability to use these models without that massive investment early in the project.

Rather than building your own data model, the Cognitive Services models are pre-trained and require no coding on the part of the developer. The Cognitive Services APIs are exposed via URL and are expected to result in a JSON payload. Figure 1-1 shows possible sources of calls to the Cognitive Services APIs (on the left) and the API destination (on the right) returning the JSON response.

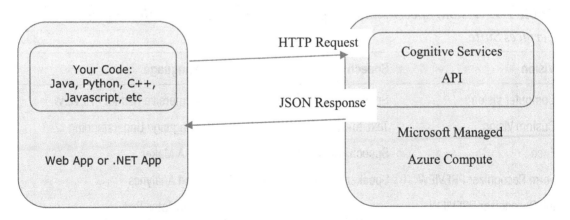

Figure 1-1. *Simplicity of the Azure Cognitive Services architecture, APIs easily referenced via code*

Cognitive Services Suite

The Microsoft Cognitive Services suite provides a collection of APIs abstracting the machine learning models for Vision, Language, Speech, Decision, and Search capabilities. The following list gives a brief description of each part of the suite:

- *Vision* – Recognize, identify, caption, index, and moderate your pictures, videos, and digital ink content.

- *Language* – Allow your apps to process natural language with prebuilt scripts, evaluate sentiment, and learn how to recognize what users want.

- *Speech* – Convert speech into text and text into natural-sounding speech. Translate from one language to another and enable speaker verification and recognition.

- *Decision* – Build apps that surface recommendations for informed and efficient decision making.

- *Search* – Add Bing search APIs to your apps and harness the ability to comb billions of web pages, images, videos, and news with a single API call.

We will be providing an overview of the full Cognitive Services suite. We will also provide real-world examples of the most commonly used features of the Vision, Speech, and Language components listed in Table 1-1.

Table 1-1. *Vision, Speech, and Language APIs Are the Base of the Cognitive Services Suite*

Vision	Speech	Language
Computer Vision	Speech-to-Text	Immersive Reader PREVIEW
Custom Vision	Text-to-Speech	Language Understanding
Face	Speech Translation	QnA Maker
Form Recognizer PREVIEW	Speaker Recognition PREVIEW	Text Analytics
Ink Recognizer PREVIEW		Translator Text
Video Indexer		

Examples of Cognitive Services at Work

Now that you have a basic understanding of what Azure Cognitive Services is at its core, what can we do with the services? While every API within the suite certainly has its specific application, we have been fortunate enough to see these at work. What follows are some real-world examples of companies and people using Azure Cognitive Services.

Flood Recognition Using the Vision API

Our first scenario will involve the Vision API and its use in disaster response (specifically, flood response). When dealing with a natural disaster, the speed of the response is critical. For things like hurricanes and tornadoes, you generally know, to a certain degree of precision, where the affected area will be, the scope of the response needed, and the type of damage you will need to assess and handle. What if the natural disaster is a flood created by unexpectedly high amounts of rain rather than a massive storm for which you have had several days of notice and preparation? What if the flood starts in a small creek or watershed? Many governments have placed cameras in flood-prone areas, and that is a good first technological step, but it is impractical to assume that you can have personnel constantly reviewing the feeds from hundreds or thousands of cameras across a region. This is where the Cognitive Services Vision API comes into play. By utilizing the Vision API to train into the model an understanding of what areas look like when they are flooded vs. when they are not flooded, you can then trigger alerts to the proper people when your application (using the Vision API) detects a potentially flooded image in a feed from one of the cameras. Leveraging the Vision API for this purpose dramatically speeds up response times, which can help ensure that lives and property are saved and the damage is mitigated.

Analysis of Social Media Posts to Alert for Aggressive Behavior Using the Language API

What if the threat is not a natural disaster but potentially a human-created disaster scenario? Unfortunately, as those in the United States have become all too accustomed to, mass shootings are an all-too-common occurrence in modern life. Many, if not most, of these mass shootings follow a familiar pattern: the potential perpetrator displays aggressive behavior or language and ends up on some sort of watch list. If they are a student, this watch list is often kept at the school or district level. If they are an adult,

5

they may be referred to mental health professionals at their place of work or within their local public health infrastructure. Often the aggressive language or behavior is displayed on social media, and the heartbreaking stories written after a violent event takes place seem to always reflect back on a pattern of social media posts indicating the person's increasing depression, declining mental state, etc. If you can cross-reference people on these watch lists with their social media posts, you may be able to see for yourself as their posts take a dark turn and potentially predict, or directly state, that they intend to commit a violent act. It's not reasonable, though, to assign a mental health professional or any school, corporate, or public resource the job of sorting through hundreds or thousands of people's social media feeds for any sign of a descent into madness.

This is where the Text Analytics API within the Language portion of the Azure Cognitive Services suite comes into play – specifically the sentiment analysis abilities. Sentiment analysis, at its most basic level, analyzes a piece of text and returns a numeric score indicating how positive or negative it believes the piece of text to be. Typically, this is represented by a decimal between 0 and 1 – 1 being very positive and 0 being very negative. Using the Text Analytics API and a bespoke application (or something like Azure Logic Apps, which will be discussed later in this book), you can analyze a person's public tweets or Facebook statuses or Instagram comments. From that analysis, you can get a baseline of the average sentiment score of that person's posts, which would allow you to know when the posts deviate far from that average into very negative territory. Based on that quantitative information, you could then take suitable action such as initiating a wellness visit, contacting parents or guardians, or even contacting the authorities. In fact, this could all be automated. There are school districts already doing this sort of thing to mitigate violence in their communities, and the Language API within Azure Cognitive Services puts that power in our hands as well.

Social Media Marketing Success Analysis Using Cognitive Services Language

Our third scenario puts us in less legally and morally tricky territory but refers again to the Language portion of the Cognitive Services suite. As most are well aware, a company's presence on social media can often have a real impact on sales of their product and, relatedly, the overall financial health of the company itself. Whether you are trying to mitigate a marketing misstep or just gauge the reaction to your latest product or service, it can be very difficult to precisely determine how current (and potential) customers and users are feeling about and interacting with your brand. We can again leverage the Language API to give us this type of data.

Has your marketing team rolled out a new campaign and the hashtag to go with it but you have no idea whether or not it's actually being received well at all? Feed all social media posts and comments into an application calling the API and let the score supplied by the sentiment analysis API tell you whether or not the new marketing campaign is engaging with customers. You now have quantitative data you can use to measure the relative success of a marketing campaign so you no longer need to rely on anecdotal evidence and gut feel. Are you a small business whose every interaction on social media goes a long a way toward determining the health of your business? Through your own development or using other Azure services, you can create a workflow where a negative sentiment score triggers a notification to you or your customer service team so they can reach out directly to the complaining customer to investigate the issue and provide a solution.

Translating Written Customer Communication to Multiple Languages Using Cognitive Services Language

In any business, communication with customers is key. In this increasingly globalized world, a business often has to interact with suppliers, employees, and customers from many different countries who, of course, speak many different languages. Localization efforts to ensure language on websites, on social media accounts, in documentation, etc. are time-consuming, and that's assuming that you can find the native speakers you need to correctly translate the source's original language into the language needed. Imagine the effort to change external and internal websites, documentation, and customer communication for just one different language! Multiply that effort by 5, 10, 20, or more, and you are starting to see the situation that many companies find themselves in with no clear path to doing it correctly. The Translator Text API within the Azure Cognitive Services Language API drastically decreases the amount of effort needed to initiate and implement localization efforts like I have described in the preceding. This API can translate to and from many languages (even Klingon for those Trekkies out there!) using simple API calls and relatively basic programming concepts to handle the response that is returned. Translation, however, may not resolve every localization issue you have.

Let's say you need to translate between languages that do not have a common alphabet – what can you do? The Translator Text API is also capable of transliteration. While many are familiar with the meaning of translation, some may not be familiar with the term transliteration. Transliteration means to "to change (letters, words, etc.) into corresponding characters of another alphabet or language." For example, "Hello, how are you?" translated into Hindi appears as "नमस्ते आप कैसे हैं?" Notably, Hindi is represented in a different alphabet than English. Having worked on software over the years that required extensive localization, transliteration efforts were often an immense challenge, especially if the company is not located in an area with a wide diversity of languages. Leveraging the Language API for these efforts dramatically cuts both the time needed to complete them and the overall cost of the effort itself.

Translating Speech in a Public Health Call Center

Finally, and in a similar vein, the Speech API within the Cognitive Services suite does for the spoken word what the Language API does for the written word. While it is capable of more passé applications such as text-to-speech or speech-to-text, as well, Speech Translation within the Speech API allows you to quickly translate spoken words into words spoken in many other languages using the structure of API calls and response handling as the other APIs within Azure Cognitive Services. Imagine that you are a public health call center in a major urban area with a great diversity of spoken languages. For example, it is estimated that around 800 languages are spoken within New York City. As a public health call center, your responsibility is to get essential public health information disseminated widely as quickly as possible. It is likely, though, that many of your employees speak only English or Spanish and maybe a handful of other languages. How can your radio and television advertisements and interactive voice response (IVR) systems serve the needs of as many citizens as possible? The information needs to be conveyed to constituents in their native language, but it is not reasonable to expect your staff to learn new languages just to do their jobs. In a situation like this, the flexibility of the Speech Translation abilities within the Speech API can be a game changer in your public health agency's ability to interact with its community.

While these five scenarios sound interesting and beneficial, there is no such thing as a free lunch, is there? That may be true, but the pricing of API calls to the Cognitive Services suite is moderate and accessible to many. Pricing details as of publication time are available in the next section.

Cognitive Services Pricing

Are you price sensitive and would like to start using the Cognitive Services APIs for educational or test/dev purposes? Then you are in luck! To make AI accessible to everyone, many of the Cognitive Services have a tier offering limited free services.

These services roll over monthly, so if you do happen to hit your limit, you can wait for the end of the month for your limit to reset. Table 1-2 shows the free tier names and capacity limits of various Cognitive Services at the time of publication.

Table 1-2. *Free Tier Services and Capacity Limits*

AI Service	Functionality	Capacity
Computer Vision	Extract rich information from images to categorize and process visual data.	5,000 transactions S1 tier
Personalizer	Deliver rich, personalized experiences for every user.	50,000 transactions S0 tier
Translator Text	Add real-time, multilanguage text translation to your apps, websites, and tools.	2,000,000 characters S0 tier
Anomaly Detector	Detect anomalies in data to quickly identify and troubleshoot issues.	20,000 transactions S0 tier
Form Recognizer	Automate the extraction of text, key/value pairs, and tables from your documents.	500 pages S0 tier
Content Moderator	Moderate text and images to provide a safer, more positive user experience.	10,000 transactions S0 tier
Custom Vision	Easily customize Computer Vision models for your unique use case.	10,000 predictions S0 tier
Face	Detect and identify people and emotions in images.	30,000 transactions S0 tier

(*continued*)

Table 1-2. (*continued*)

AI Service	Functionality	Capacity
Ink Recognizer	Recognize digital ink content, such as handwriting, shapes, and document layout.	2,000 transactions S0 tier
Language Understanding	Build natural language understanding into apps, bots, and IoT devices.	10,000 text request transactions S0 tier
QnA Maker	Create a conversational question-and-answer bot from your existing content.	3 days S0 tier
Text Analytics	Extract information such as sentiment, key phrases, named entities, and language from your text.	5,000 transactions S tier

Azure Cognitive Services Architecture

Now that we know what the Cognitive Services can do, let's talk about how they work. What are these services? What components make up the services and APIs? How do we interact with them? What infrastructure and/or code do we have to manage?

Lambda Architecture Overview

You can integrate data and AI capabilities by leveraging a lambda architecture in application design. A lambda architecture is a data processing architecture designed to process big data in batch and speed latency scenarios. For training purposes, machine learning models require access to bulk historical data. A lambda architecture provides a structure for data to be consumed by machine learning models for model training. It also provides a data flow for routing live data to the machine learning models for analyzing, classifying, and predicting outcomes.

A lambda architecture is comprised of the batch, speed, and serving layers. The batch layer represents a traditional immutable master dataset, similar to a traditional operational data store or enterprise data warehouse. Data within the batch layer is consolidated and pre-aggregated into batch views. These views are exposed to consumers for reporting in the serving layer. Lastly, the speed layer serves data more quickly, by providing more instant access to only recent data via real-time views. In a lambda architecture, consumers can query batch views and real-time views or query and merge results from both.

A lambda architecture can be easily leveraged by AI systems. Organizations store historical data in a multi-aggregated form. These data points stored in the batch layer can be used to train a data model via the serving layer. Through application modernization, new data can be brought in through the speed layer and immediately processed against a machine learning model. The results can then be stored and immediately leveraged for business insights. Figure 1-2 shows a reference architecture utilizing Microsoft data and AI resources and adhering to lambda architecture design.

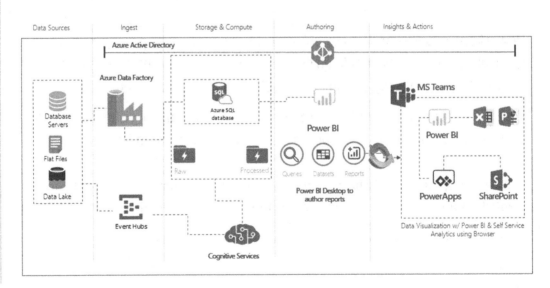

Figure 1-2. *Data + AI solution architecture*

As you can see, the preceding architecture depicts how we are able to integrate Cognitive Services in an existing reporting/transactional system. Most enterprises will already have a batch layer in place. They will have an operational data store or enterprise data warehouse that serves as a data repository. The standard process to update these repositories with the latest data tends to be ETL (Extract, Transform, and Load) jobs that run hourly or, most often than not, nightly. We can leverage these curated datasets for training and optimization of our Cognitive Services models while taking advantage of capabilities like Event Hubs to perform real-time classification and labeling by passing data via the speed layer. Finally, after this information is written to the central repository, it is exposed to data consumers via the serving layer. We can also easily integrate with existing application systems.

Embed AI Speech + Text + Vision in an Application

There are a number of ways to embed Cognitive Services Speech, Text, and/or Vision within an application. Since the Cognitive Services are API based, there are a number of languages that are easily able to interact with the API services. In addition to developers accessing the APIs with code, analysts and business users can leverage AI capabilities directly via Microsoft Office applications such as PowerApps and Power BI via Teams.

PowerApps and Power BI   PowerApps (`https://powerapps.microsoft.com`) is a tool allowing developers to quickly build apps using pre-built templates, drag-and-drop simplicity, and integrated CI/CD enabling quick deployment.

Power BI (`https://powerbi.microsoft.com`) is a reporting and analytics tool allowing for advanced ad hoc reporting and enhanced visualization.

It's exciting stuff! We will have examples for you in Chapter 6! Figure 1-3 depicts a potential solution architecture for an application embedding AI Speech + Text + Vision.

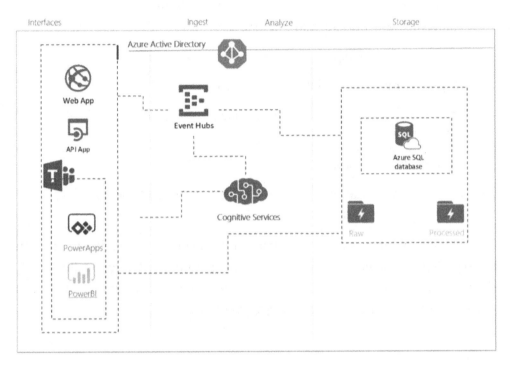

Solution Architecture
Application Development - Embed AI Speech + Text + Vision

Figure 1-3. *Embedded solution architecture*

While the preceding reference diagrams detail common use cases, they absolutely do not detail *all* the possible ways of interweaving AI into your technology stack! We will be providing many other examples in subsequent chapters.

API References

The predetermined parameters and results of the Cognitive Services APIs are detailed within API Reference documentation available within the Cognitive Services documentation pages. Each API has a Reference document detailing the following:

- *Region availability* – Cognitive Services APIs are not available in all regions. As managed services, with code + compute + storage maintained by Microsoft, the APIs are available and maintained in select regions.

13

- *Overview of the API call* – Brief description of API expected results.

- *Format of the request URL* – Example of URL formatting to interact with the API.

- *API call parameters* – Expected parameters including value type, valid types, and optional parameters.

- *Request headers* – Expected Header values in a message.

- *Request body* – Expected Body values in a message.

- *Error responses* – Expected error message, that is, error types 200, 400, 415, and 500.

- *Example JSON payload* – Example of API results.

- *Code samples* – Examples of calling the API using Curl, C#, Java, JavaScript, ObjC, PHP, Python, and Ruby.

The code samples are code snippets that will show you sample code with the details needed to pass your headers, subscription key, and parameters to the specific API described. You will notice that most API calls can be made with fewer than 20 lines of code! Figure 1-4 shows part of constructing an API call in code.

Figure 1-4. *A partial sample of an API call*

Getting Started

At this point, you must be wondering... "How do I get started using the Cognitive Services APIs?" You will need an Azure subscription and some lightweight tools to start writing code. As of the time of this writing, you can sign up for an Azure free account that provides 12 months of access to free services.

Now that you have access to the APIs, you will need a method to interact with them. Two lightweight and free code editors are Azure Notebooks and Visual Studio (VS) Code.

Azure Notebooks

In the initial drafts of this book, you could find a preview Azure Notebooks experience at notebooks.azure.com. As work on the book continued, the preview experience ended. Currently, Microsoft directs those interested in notebook experiences with Cognitive Services to use notebooks in Visual Studio Code (explained in further detail in the next section) and GitHub Codespaces (which is currently in beta at the time of publication). Codespaces allows you to edit your notebooks using VS Code or your web browser and store them on GitHub.

VS Code

VS Code is a source code editor developed by Microsoft; it is not an Integrated Development Environment (IDE). An Integrated Development Environment is a tool that lets you write, compile, and debug code. VS Code is an editor that provides support for debugging via break points, call stacks, and an interactive console. It allows for embedding with Git and other SCM providers for source control, enabling the review of diffs, source code commits, and version push and pull directly, as well as syntax highlighting, intelligent code completion, snippets, and code refactoring via Microsoft's IntelliSense.

Cognitive Services SDKs

Microsoft offers a collection of Cognitive Services Software Development Kits (SDKs) in the languages C#, Java, Python, Go, JavaScript, and R. Within the API Reference pages for the Cognitive Services APIs, you will find code snippets in various languages to help you interact with the APIs.

Learning Resources

The goal of this book is to be a resource to help you get up and running with AI Cognitive Services in a very short period of time. To help you with your journey, we would also like to point you to alternate available online resources. While none of these resources are prerequisites for this book, we would like you to benefit from opportunities to supplement your learning with video tutorials and hands-on labs and, when you are ready, advanced concepts.

Microsoft Learn

Microsoft Learn is an online education platform with material curated and hosted by Microsoft. Coursework is comprised of modules aligned with learning paths organized around roles (developer, architect, system administrator) or technologies (i.e., web apps, Power BI, Xamarin).

There are a number of AI Cognitive Services modules currently available:

- Add conversational intelligence to your apps by using Language Understanding Intelligent Service (LUIS).

- Discover sentiment in text with the Text Analytics API.

- Evaluate text with Azure Language Cognitive Services.

- Recognize specific voices with the Speaker Recognition API in Azure Cognitive Services.

- Classify and moderate text with Azure Content Moderator.

- Process images with the Computer Vision service.

- Translate speech in real time with Azure Cognitive Services.

- Classify images with the Microsoft Custom Vision service.

In addition, there are a few AI Cognitive Services learning paths as well:

- Process and translate speech with Azure Speech Cognitive Services.

- Process and classify images with Azure Vision Cognitive Services.

Summary

In this chapter, you learned the following:

- The Microsoft Cognitive Services APIs are pre-built and require no machine learning programming to deploy.

- A lambda architecture can be employed to train a Cognitive Services model using historical data and provide predictions via a real-time analytics data flow.

- APIs allow developers to interact with Cognitive Services models using multiple languages such as C#, Java, JavaScript, ObjC, PHP, Python, and Ruby.

- It is easy to get started. There are free tiers for Cognitive Services and free applications such as Visual Studio Code or GitHub Codespaces for code authoring.

We hope that you have enough information to understand what the Cognitive Services APIs are and on how to find additional sources of information for furthering your learnings. For the rest of the book, we will focus on examples of leveraging Cognitive Services APIs taken directly from the field. Feel free to skip around or through the book. Each chapter is meant to be self-contained and holds enough detail for you to execute examples start to finish without having to rely on material from other chapters.

CHAPTER 2

Prerequisites and Tools

While the Cognitive Services APIs enable developers and data professionals to put a lot of cool technology to work, it's important to talk about the tools used to develop, test, and operationalize this technology. Before one can dive into the tools, it's important to talk about the prerequisites necessary to working with the Cognitive Services APIs in development environments, test environments, and production environments. Before we can discuss prerequisites or tools, however, it is important to lay out an intro to Cognitive Services and Machine Learning, which programming languages lend themselves to the best (and easiest) interaction with these APIs. While there are some languages whose interactions with these APIs may best be described with the terms "square peg" and "round hole," we will walk you through the most common interactions with the Cognitive Services APIs in some frequently seen and used programming languages.

Cognitive Services and Machine Learning

Microsoft created Cognitive Services both for developers without machine learning experience and to assist experienced data scientists. Cognitive Services provides elements for both audiences. As the Cognitive Services are APIs, these services and their SDKs provide prebuilt AI models to be leveraged in intelligent applications as well as allow for their use in experiments. The Cognitive Services were designed to provide solutions to complex analysis tasks including sentiment analysis, object detection, face detection, search, or image classification. The Cognitive Services include accelerators for testing use cases and allow anyone to utilize thousands of man-hours and research used to create these complex libraries, allowing people to add this complex logic easily in models without having to create and train a model from scratch. Naturally Cognitive Services can be part of solutions created in Azure Machine Learning, Azure Synapse, and Databricks Machine Learning, which are tailored toward data scientists with machine

© Alicia Moniz, Matt Gordon, Ida Bergum, Mia Chang, Ginger Grant 2021
A. Moniz et al., *Beginning Azure Cognitive Services*, https://doi.org/10.1007/978-1-4842-7176-6_2

learning expertise. In addition to these applications, Microsoft created an open source library MML to assist Spark MLlib users to utilize the functionality provided by Cognitive Services. The Cognitive Services SDKs can also be used in Azure ML, Azure Synapse, or Azure Databricks notebooks and allow you to run large workloads and distribute data without having to worry about the underlying infrastructure. Machine learning generally is used to solve a problem where there is a lot of data that can be used to train an algorithm to learn how to solve the problem for a given use case. The process used involves splitting the data into test and training datasets and using algorithms to determine the solution. An understanding of these concepts will be beneficial in learning Cognitive Services. Before you start developing, we will introduce you to some concepts in Machine Learning that we believe will be beneficial.

In Machine Learning, there are different categories and models meant for solving different problems. First of all, we have supervised and unsupervised learning. Supervised learning is when training data comes with a description, target, or desired output and we find a rule to map inputs to outputs and to label new data. Unsupervised learning is when training data doesn't contain any description or labeling; that's when it's up to us or the service to uncover hidden insights. Clustering and Anomaly Detection falls into the unsupervised category.

Supervised learning can be divided into three categories:

- Binary classification or prediction classifies the elements of two given sets into two groups and predicts which group each belongs to. For example a student can "pass" or "fail";, a patient can test positive or negative for disease or a comment can be classified as "positive" and "negative" for sentiment analysis.

- General classification or multi-class classification refers to a predictive modeling problem with more than two class labels. Unlike binary classification, it does not have the notion of normal and abnormal outcomes. Instead something is classified to belong to one among a range of known classes. This is where more of the Cognitive Services models apply. The Image classification Cognitive Service falls into this category where the response describes the image content and attributes – for example, fruit/banana or fruit/mango in a fruit dish, where the model predicts a photo to belong to one among thousands of fruits.

- Regression outputs a predicted numerical value – for example, to predict house prices.

There are different algorithms below each of these, with advantages and disadvantages, but we'll stop here. If interested, there is a good primer on machine learning algorithms available here: `https://azure.microsoft.com/en-us/overview/machine-learning-algorithms/`. Additionally, there are many books deep diving into these topics that are present on Amazon's Best Sellers in AI & Machine Learning list: `www.amazon.com/Best-Sellers-Books-AI-Machine-Learning/zgbs/books/3887`.

Now that you have a basic understanding of what AI capabilities are at their core and what they mean, how should we start using them in our solutions, and what tools do we need to get started?

The AI Engineer Toolkit

To start a journey to become an AI engineer, there are various skillsets that your future colleagues or client might expect from you. These skillsets include both tech and non-tech ones, from data handling and data exploration to machine learning modeling and from developing a model to shipping the model to a product.

Every journey begins with data! Two tools we enjoyed having when we started our AI engineer/data scientist journeys: Azure Data Studio and SQL Server Management Studio (SSMS). Many of us are already using SSMS, but we would like to introduce you to the fairly new Microsoft tool, Azure Data Studio.

Azure Data Studio

The tool integrates well with SQL Server, Azure SQL Database, Azure SQL Data Warehouse, and PostgreSQL on Azure and works on Linux and Window operating systems. The download link can be accessed from here: `https://docs.microsoft.com/en-us/sql/azure-data-studio/download-azure-data-studio`.

Azure Data Studio was previously released under the name SQL Operations Studio. Then it was moved to general availability (GA) on September 24, 2018. Figure 2-1 shows a view of the application.

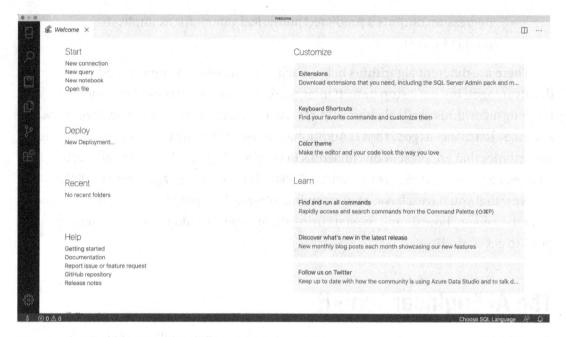

Figure 2-1. *Azure Data Studio interface*

Many developers compared it with SQL Server Management Studio. Both tools provide the functionality to work with MS SQL Server and Azure products really well. But Azure Data Studio includes more functionality for data engineering and data visualization than SMSS. Table 2-1 gives a brief overview of the pros and cons of these two services.

Table 2-1. *SSMS vs. Azure Data Studio*

	SSMS	Azure Data Studio
Features	– Supports Windows only	– Supports Windows, macOS, Linux
	– Table designer	– Multiple tab windows
	– Database diagram tool	– SQL IntelliSense
	– Error log viewer	– Supports Jupyter Notebook
	– Import/export DACPAC	– Source control integration (Git)
	– Script wizard	– Quickly charts and visualizes result sets
	– Performance tuning advisors	– Exports data view to CSV, JSON, or Excel easily
	– Maintenance plans	

(continued)

Table 2-1. (*continued*)

	SSMS	Azure Data Studio
Target user	– Windows user – Database administration role – CRUDs the database often – Doing more tasks regarding security management	– macOS or Linux user – Data engineer or AI engineer role – Focus on ed\iting or executing queries more than administrative features – Requires data visualizations of the SQL result quickly

To summarize all of the comparisons, if you are a cross-platform developer or work as an AI engineer more than a database administrator, then we would definitely recommend you try Azure Data Studio.

Machine Learning or Deep Learning Frameworks

After exploring the data, the next step will be working on the machine learning model. For the typical machine learning use case, Azure Cognitive Services provides you a pre-trained model that works quite well. There will be some cases where you will need to customize the model or add extra logic into your application, to make the prediction of the pre-trained model follow the business scenario that you are working on, that is, the scenario where you will have to work with a machine learning or a deep learning framework. The following sections describe some frameworks that have been widely used in the data science area.

Pandas

Pandas (`https://pypi.org/project/pandas/`) is an open source Python library that is built on top of NumPy, which provides a data structure called a DataFrame. A DataFrame enables you to work with table data with the same mindset of manipulating the data with SQL or Excel. Between tabular data you retrieved from the database to the machine learning model, Pandas will help you perform the data cleaning operations, such as inserting or deleting data, grouping data by the given statement, merging datasets, etc.

Scikit-Learn

Scikit-Learn (`https://pypi.org/project/scikit-learn/`) is a Python library for machine learning built on top of SciPy, which is one of the most popular frameworks for building machine learning models. In Scikit-Learn, you can use classification, regression, and clustering models. You can also perform model optimization with dimensionality reduction and model selection with grid search and metrics. Eighty percent of AI questions are really machine learning questions, and you can probably use Scikit-Learn to solve them.

Spark MLlib

Many Machine Learning solutions are now being developed with Spark MLlib (`https://spark.apache.org/mllib/`), an open source framework, as the scalability of the underlying structure allows for creating solutions in a decreased time than other development libraries. In order to use Spark, machine learning components must be written using its MLlib library, which contains a number of popular algorithms and has the ability to call Tensor frames. It includes Scikit-Learn, but is incompatible with Pandas as its resilient DataFrames must be used to store data so that it can be accessed at scale. It provides an alternative to Pandas, the Koalas library, to provide the same functionality as Pandas. It supports a number of different languages including C#, Spark, R, Java, and Scala.

TensorFlow

TensorFlow (`https://pypi.org/project/tensorflow/`) is an open source software library for high-performance numerical computation. The name Tensor comes from the data TensorFlow uses to compute. A Tensor can be a variable, representing a constant, a vector, or a matrix. Tensors usually refer to high-dimensional matrices. Table 2-2 shows some examples to help you see what a Tensor looks like.

Table 2-2. *Tensor Examples*

Name	Dimension	Example
Scalar	0D Tensor	3
Vector	1D Tensor	[-1,2,7,9]
Matrix	2D Tensor	[[-1,2,7,9], [5, -2,5, -3]]
Matrices	3D Tensor or higher-dimensional Tensor	[[[-1,2,7,9], [5, -2,5, -3]], [[7,8,31, -4], [8,10, -45, -73]]]

TensorFlow is a great choice if you are interested in customizing a deep learning model and you want to deploy the model to desktops, clusters of servers, mobile devices, or even edge devices.

CNTK

The Microsoft CNTK (Cognitive Toolkit, `https://pypi.org/project/cntk/`) is a free, easy-to-use, open source, commercial-grade toolkit that trains deep learning algorithms to learn like the human brain. CNTK can be used in Python, C#, or C++ programs and provides similar functionality as TensorFlow. You can use ONNX to export a CNTK model as a TensorFlow, PyTorch, Keras, or MXNet model.

These are the machine learning frameworks. Though they are outside of the scope of this book, after you have worked with the Cognitive Services, I would recommend you pick up at least one or two of them. After performing your initial use case and project with a Cognitive Service, try building a model from scratch with the same business requirements. After working with Cognitive Services, you may notice an improvement in your ability to work with machine learning models at a higher level of control, for example, knowing how to measure the model with proper metrics and knowing how to compare different parameter setups.

Azure AI Gallery

The previous sections reviewed different data exploration and modeling tools and have provided the background to delve deeper into what a machine learning project can achieve and what kind of problems it can solve. Figure 2-2 shows Azure AI Gallery (`https://gallery.azure.ai/`), which you can visit to get more ideas around Azure AI use cases and projects.

Figure 2-2. *Azure AI Gallery*

The Azure AI Gallery is a fun way to browse sample projects and get familiar with solutions for common business problems. The AI Gallery provides common use cases categorized by industries, such as retail, manufacturing, banking, and healthcare. Take retail as an example. It contains the study cases of demand forecasting and price optimization. You can visit either case to know what a typical dataset of this problem looks like, what features are important in this case, or even which algorithm will be the best fit in this case.

After reviewing, you will have more experience both practical and theoretical on machine learning projects. You will also have more domain know-how and intuition of how an AI project should look like.

Programming Languages

The AI solutions covered in this book use the Microsoft Cognitive Services APIs and SDKs to leverage complex libraries created to analyze image content, detect and identify people, identify forms and extract selected data, create bots, translate text, and many more. These APIs are designed to decrease the time it takes to develop solutions without needing to have experience in Machine Learning. You will not need to know how to code a machine learning module from scratch, but you will need to develop code to pass variables containing the text, audio, or image file that you are analyzing to the

API. And of course, you will have to develop code to receive the results. Fortunately, there is a lot of sample code that exists to assist in creating the solutions. Microsoft has designed these solutions to be implemented mainly in two languages, C# and Python; however, examples are also readily available in Go, Java, and JavaScript. SQL is used to supplement these languages by providing a method of accessing the data to be analyzed for training, testing, and implementing solutions. Let's review how these languages can be used to implement Cognitive Services.

C#

C# is written and supported by Microsoft. It was created as a flexible object-oriented language, which can be used to create a number of different applications for many different environments, including client applications, web services, and mobile applications, and to call Azure services. C# was modeled after C++ and Java, and people who are familiar with those two languages will be comfortable in C# in a short time. C# contains two underlying components that provide the language the ability to use prebuilt objects allowing the developers to rely on libraries as well as support for underlying hardware, which allows it to be run on any machine. It uses the .NET Framework and .NET Core to store the shared code used in the language called the Framework Class Library. .NET Core will run on Windows, Linux, and MacOS. The environment for applications, called the Common Language Runtime (CLR), will manage memory and processor threads to abstract the application from the hardware. In this book, we will be providing a number of samples using C# to illustrate how to call Cognitive Services APIs.

When writing C# code, you can use a number of different Integrated Development Environments (IDEs) including Visual Studio Code (VS Code), Visual Studio, Visual Studio Code Online, or Jupyter Notebook. All of these tools have free versions that you can use to develop your code. You will need to install the client library for the Vision, Language, and Speech APIs discussed in this book. In Visual Studio, you will add these by accessing the **Manage NuGet Packages** option and browsing for the API you wish to load.

Python

Python is an open source language administered by the Python Software Foundation. A script-based object-oriented language, Python is arguably one of the most popular languages for all kinds of different development tasks and interactive and noninteractive coding tasks on a variety of different platforms. Its strength relies on a vast number of widely available libraries that provide it the ability to form a number of different libraries. Python is used to create a variety of different applications including complex websites and to create machine learning and AI solutions, as well as for mobile phone development. Python code is developed most often using virtual environments, which allow you to encapsulate different versions of Python and libraries into one area for a given project. You will want to create virtual environments for the code created in this course. The Python code used in this course was created with version 3.8, the latest version as of this writing.

Python supports a number of different UIs including VS Code, Visual Studio, Visual Studio Code Online, Atom, Sypder, PyCharm, and Jupyter Notebook to name a few. There are many different paid and free UIs for you to choose. Feel free to select the UI that you wish to use. You will need to install the dependencies with pip for Cognitive Services as directed in later parts of this book.

SQL

SQL is the most common language used for gathering and managing data. Originally created for relational databases, SQL is now used to access data stored in many different places, including files and other unstructured formats. Most variants of SQL base their version on the current standard version implemented by the American National Standards Institute (ANSI). Understanding how to query and manipulate data with these commands will provide a good baseline for working with all relational databases. Many SQL variants including SQL Server's T-SQL and Oracle's PL/SQL contain additional features not included in the ANSI standard. Both Python and C# include libraries that allow SQL statements to be used within code. The SQL code is used to access and combine different data sources into datasets that can be used for machine learning training and applying models. You won't need a separate UI for writing SQL code, but you can use Jupyter Notebook, Azure Data Studio, Visual Studio, or SQL Server Management Studio if you want to query databases or, in the case of Azure Data Studio and Jupyter Notebook, for text-based data sources.

API Code Samples

How easy is it to get started using code samples from the API pages? You can navigate to the Quickstart: Use the Computer Vision client library and select the language of your choice. When prompted to "Choose a programming language of choice," you can select "Python." You can then navigate to the GitHub repository referenced, `https://github.com/Azure/azure-sdk-for-python/tree/master/sdk/cognitiveservices/azure-cognitiveservices-vision-computervision`, where you will find detailed instructions and sample code to perform the following tasks:

- Create a Cognitive Services API.

- Install the Cognitive Services Computer Vision SDK.

- Authenticate and define variables that hold the resource group and account name associated to the Cognitive Services resource.

- Create a Cognitive Services client.

Additionally, detailed code snippets are included as examples on how to perform the following tasks:

- Analyze an image.

- Get a subject domain list.

- Analyze an image by domain.

- Get text description of an image.

- Get handwritten text from an image.

- Generate thumbnail.

You can either follow along with the instructions, or if you really don't want to start from scratch, you can download sample files from `https://github.com/Azure-Samples`. For example, Figure 2-3 shows the GitHub library for the Azure Cognitive Services Computer Vision SDK for Python.

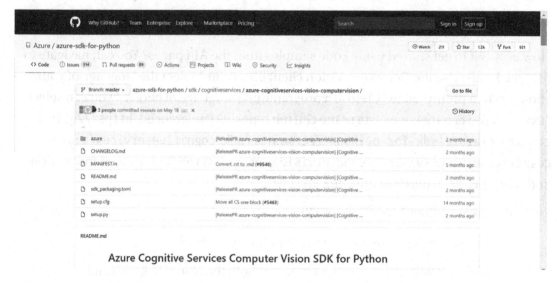

Azure Cognitive Services Computer Vision SDK for Python

Figure 2-3. *GitHub repo with Azure SDK for Python*

Integrating Code with GitHub

AI and machine learning have long been excluded from the standards that were applied
for other development projects. As the number of projects and the number of people
involved has increased, AI projects are expected to follow the same code development
processes used by other projects. For an AI project, there are two different elements
that must be defined, the source and format of the data to be used in the AI project
and the code used to create and train the AI project as well as the completed model.
Generally speaking, the data files are described using YAML Ain't Markup Language
(YAML), which can be generated in Python. YAML files are also used in Kubernetes
deployments to define the data structures. The idea is to provide continuous integration
between the code being developed and source control. While you can use any source
control, by convention we will be using GitHub. To automate this process, you may be
interested in implementing Azure Pipelines as it provides a user interface for creating
and maintaining the integration with GitHub and Azure DevOps. For more information
on implementing Azure Pipelines, you might want to check out this link: `https://docs.`
`microsoft.com/en-us/azure/devops/pipelines/?view=azure-devops`.

GitHub Repository

GitHub is a free cloud repository for storing repositories. It has versioning capabilities for all files stored within it, allowing for collaborative development as you know who made what changes when. Repositories typically contain folders containing different elements of the project including data, configuration, notebooks, etc. For more information on creating a GitHub repository, please see the GitHub guide here: `https://guides.github.com/activities/hello-world/`.

Using Markdown in GitHub README.md Files

All GitHub repositories should contain a Readme file describing the contents of the repository. This file is formatted as a Markdown file, the same format that is also used to comment in Jupyter notebooks. Markdown files can include pictures, videos, and links as well as formatted text. Markdown is simple to learn, and there are cheat sheets online to get you started if you haven't learned it already.

Defining Data and Configurations in YAML Files

The YAML file is case sensitive and should have .yml as an extension. YAML files are meant to contain lists of data items or Pipelines and are similar to JSON files in format. You cannot use tabs when creating a YAML file; everything must be in spaces. Comments in a YAML file are denoted by a hashtag #, which is also known as a pound sign. YAML indicates levels by indentation with spaces and stores items as documents. A document starts with three hyphens (---) and ends with an ellipsis (...). For example, let's define a YAML document containing information about diamonds. The file contains five columns: a carat float column, color text column, clarity text column, certificate designation text column, and price integer column. Here is how you might describe the file in YAML:

```
---#Diamonds
[carat, color, clarity, certificate, price]
...
```

YAML can also be used to contain the data as well:

```
---#Diamonds
carat: .68
color: D
clarity: VS2
certificate: GIA
Price: 5236
...
```

Downloading Sample Files from GitHub

GitHub provides the ability to check out code libraries, where a local copy is downloaded with a reference to the original files. This enables easy syncing of files, when updates are available and a follow-up checkout is triggered. Optionally, library files can be downloaded within a .zip file. You can navigate to the Quickstart code and select the green "Code" button.

There are a few ways to clone a GitHub repository. You can either pass the GitHub library path to VS Code, to GitHub, or within your Visual Studio Codespaces. Figure 2-4 shows that from GitHub online, you can obtain the following library path from the GitHub project page:

```
https://github.com/Azure-Samples/cognitive-services-quickstart-code.git
```

Figure 2-4. *Locate library URL for GitHub repo*

Figure 2-5 shows where we can add the repo URL, as well as the local path for the directory GitHub will clone the repo files into, that is, "C:\Users\alicia.moniz\ Documents\GitHub\cognitive-services-python-sdk-samples." If you would like the repo files to be cloned to a different directory, you can update the path as needed and select "Clone."

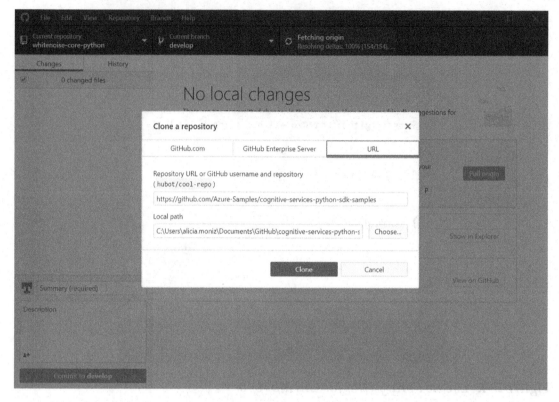

Figure 2-5. *Open GitHub Desktop and update repo URL and local path*

Alternatively, you can create a new library, called a "Repository," within your local path and extract your .zip file there. From within GitHub Desktop, you can select "Add" and then "Create new repository" as shown in Figure 2-6.

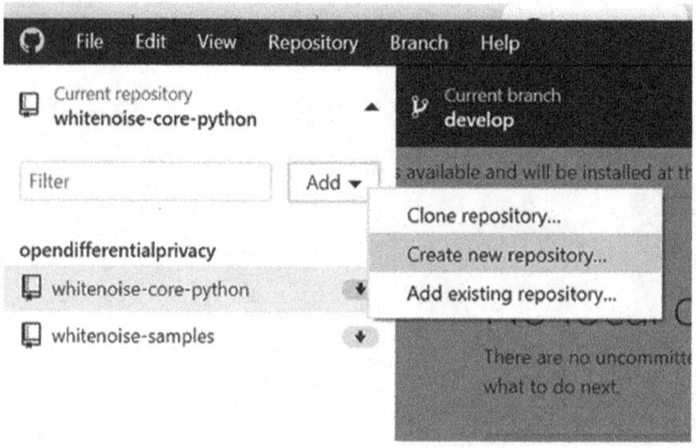

Figure 2-6. *Creating a new repository*

Again, we have the option of updating the local path to reflect the directory where we would prefer to have our files synced as shown in Figure 2-7.

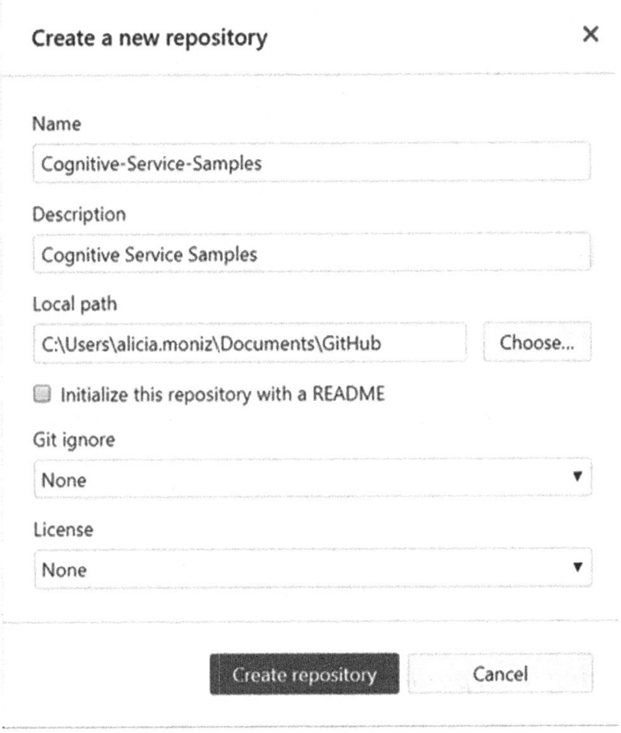

Create a new repository   ✕

Name

Cognitive-Service-Samples

Description

Cognitive Service Samples

Local path

C:\Users\alicia.moniz\Documents\GitHub   Choose...

☐ Initialize this repository with a README

Git ignore

None   ▾

License

None   ▾

Create repository   Cancel

Figure 2-7. Set local path for new repository

Next, fill in the required fields, Name and Description. Leave the Git ignore and License fields set to "None," meaning you are not excluding any files from source control syncing and you are not including a license for distribution.

Finally, click "Publish repository" as shown in Figure 2-8 to make your code available via GitHub online.

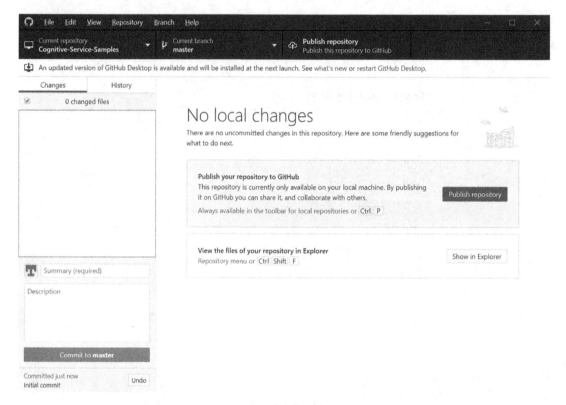

Figure 2-8. *Publishing the repository*

After you have filled in the fields, you can select "Create repository" and select Show in Explorer, to open a folder view of your repository.

At this point, as shown in Figure 2-9, your repo only has a file to hold your repository attributes, called ".gitattributes."

Figure 2-9. *A brand-new repository, synced with GitHub*

You can now drop in the .zip file that you downloaded and extract your GitHub repository. GitHub documentation on how to create and clone repositories is available.

Creating APIs via Azure Portal

You can very easily create a Cognitive Services API via Azure Portal, portal.azure.com.

After signing in to the portal, you can then search for Cognitive Services as shown in Figure 2-10.

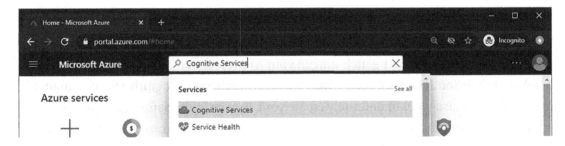

Figure 2-10. *Search results for Cognitive Services via the main portal page*

The Cognitive Services icon looks like a brain. You can quickly select "Cognitive Services" from the marketplace, as shown in Figure 2-11.

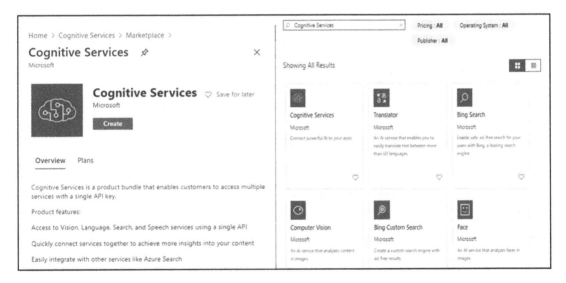

Figure 2-11. *The service we are working with is called "Cognitive Services"*

Tip   Creating a "Cognitive Services" service gives you access to all of the Cognitive Services via one service and endpoint. Instead of creating a Computer Vision service and a Face service, you can create one Cognitive Service that will provide an endpoint for both Computer Vision and Face.

You will need to provide a region, name, and pricing tier for your service. Create your Cognitive Service in the same region as your other services. For example, if you have images or data that you will be including in your AI project that are in blob storage or a database in the South Central US region, then also create your Cognitive Service in that region. You will also want to definitely create your service in the same region as your application server, if your application is hosted in Azure, to minimize response and data transfer time between the API and your application.

In Figure 2-12, we will create a Cognitive Service with the following settings: Region = "**South Central US**," Name = "**cogsvsAIusingMSFTCogSvc**," and Pricing tier = **S0**.

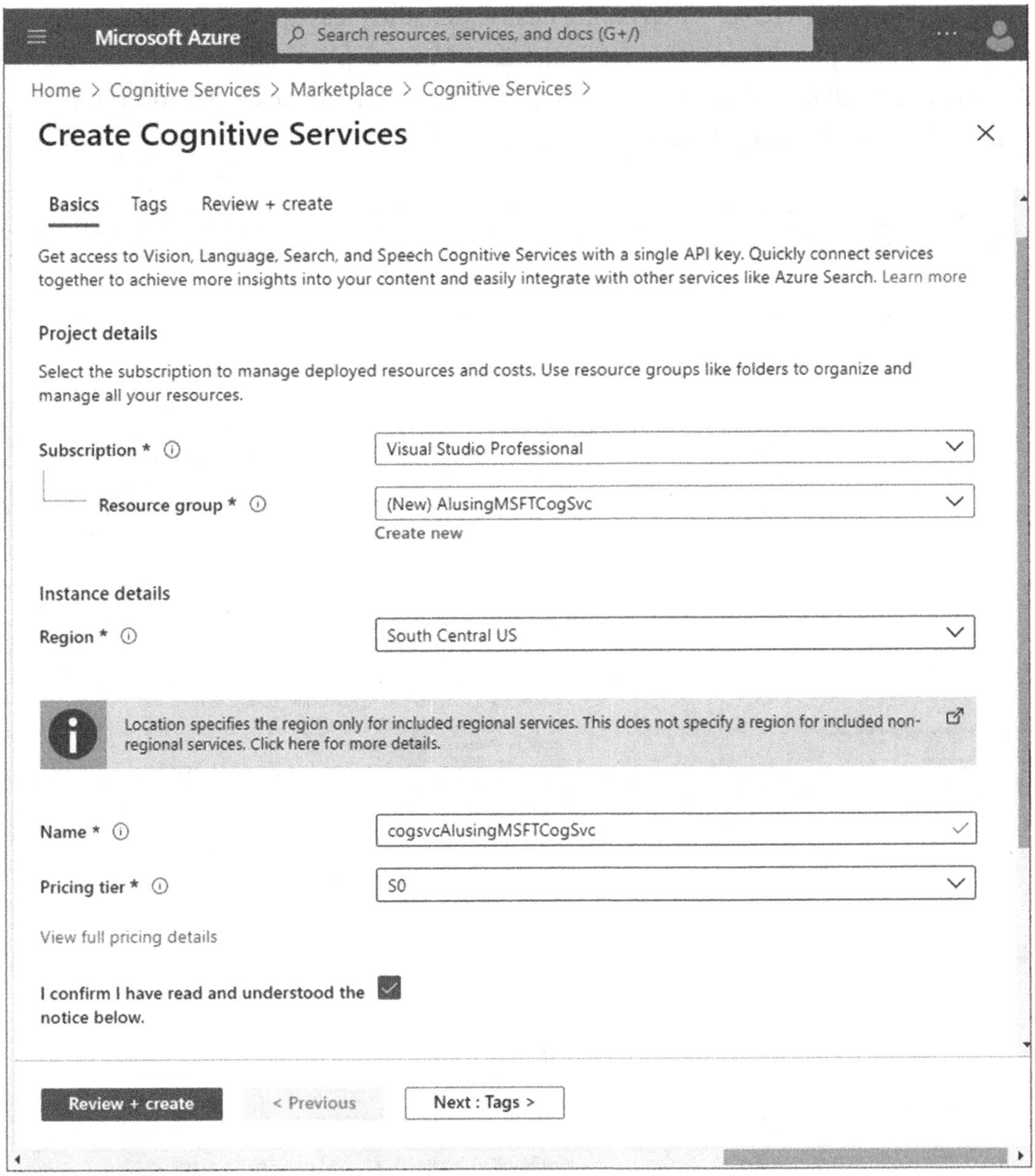

Figure 2-12. *Indicate a region, name, and pricing tier for your service*

Select "Review + create." Then on the next screen, the message "Deployment is in progress" will display. The message will change to "Your deployment is complete" after the service is available. Click "Go to Resource" to take you to the resource's overview page which is shown in Figure 2-13.

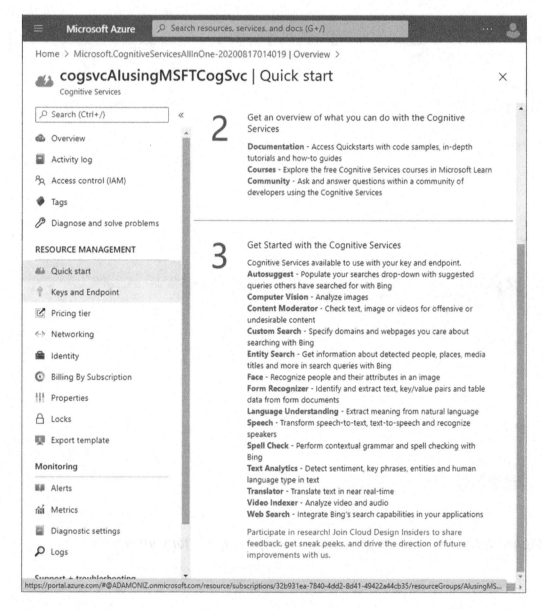

Figure 2-13. *Overview page of your new resource*

In addition to Computer Vision and Face, notice there are quite a few Cognitive Services available.

To interact with your service, you will need the endpoint and one of two keys. The endpoint is a URL, and the key will provide you access. As shown in Figure 2-14, click "Show Keys" to enable copying of your key in text format.

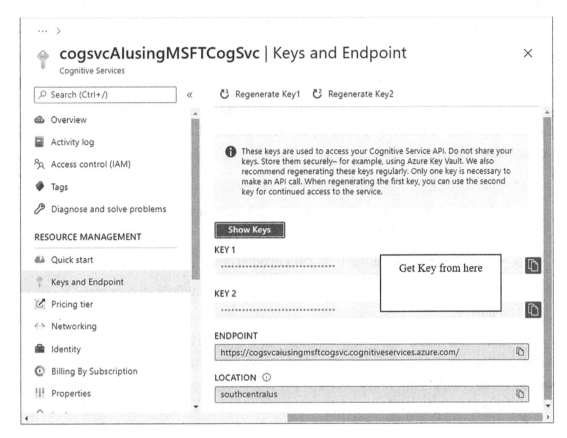

Figure 2-14. *The "Show Keys" button enables you to copy your key*

Only distribute one key to your developers. If you need to change credentials when you publish to production or have a change in staffing, distribute key #2 and regenerate key #1.

Calling APIs with Postman

Now that you have a Cognitive Services API, you need a method to interact with it easily and quickly. This also streamlines testing and debugging. The Postman API client is a helpful and easy-to-use tool. You can very easily pass text, an image file, or a URL to a Cognitive Services API with no code. This allows programmers to quickly check that the service is available and providing expected results.

The Postman API Client

The application allows users to send HTTP requests and receive and review responses. Requests can have the following body types – URL-encoded, multipart, or containing raw or binary data:

- *URL-encoded* – For sending simple text data

- *Multipart/form-data* – For sending large quantities of binary data, text with non-ASCII characters

- *Raw body editing* – For sending unencoded data

- *Binary data* – For sending media, such as image, audio, video, or text files

Postman's interface allows the developer to view metrics such as the status code, response time, and response size. The client is available for download here: `www.postman.com/product/api-client`. Downloading and installing the Postman API client is pretty straightforward. We recommend installing the application for quick testing, to simplify passing data to our Cognitive Services APIs and enable quick viewing of the JSON results returned.

Calling the Cognitive Services API with the Postman API Client

To interact with your Cognitive Services API, you will need your subscription key and URL endpoint for your API. Then using those you are able to pass a message to the API.

Find the Subscription Key and URL for Your Cognitive Services API

As shown in Figure 2-14, you can log in to your Azure Portal and navigate to your resource. On the Quick Start tab, you can copy your subscription key and endpoint URL. The endpoint URL provides you with connectivity to your API, while your subscription key will serve as your password.

The Postman API client will expect a full path for you to send form data to. From the Quick Start tab, you can click "API Console" to send you to the API console for the version of the service you have configured. Note that the page displays the full URL that needs to be referenced, which is comprised of your endpoint URL and expected parameters. Figure 2-15 shows the API console where you can identify the full URL with which to reference your Cognitive Services API.

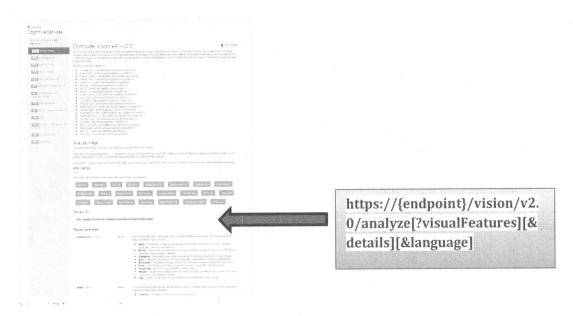

Figure 2-15. *Displays the API console and the full URL to your API*

Pass the Message to the API Using the Postman API Client

To interact with your API, you will use the Postman API client to create a POST request and send via HTTP. Within the interface, you can select "New" and then select "Request" from the available items. If you do not already have a collection, you can quickly create one. The first step in working with the application is to create a collection. As shown in Figure 2-16, within the Create Request tab, you can select "Create Collection" and key in your collection name.

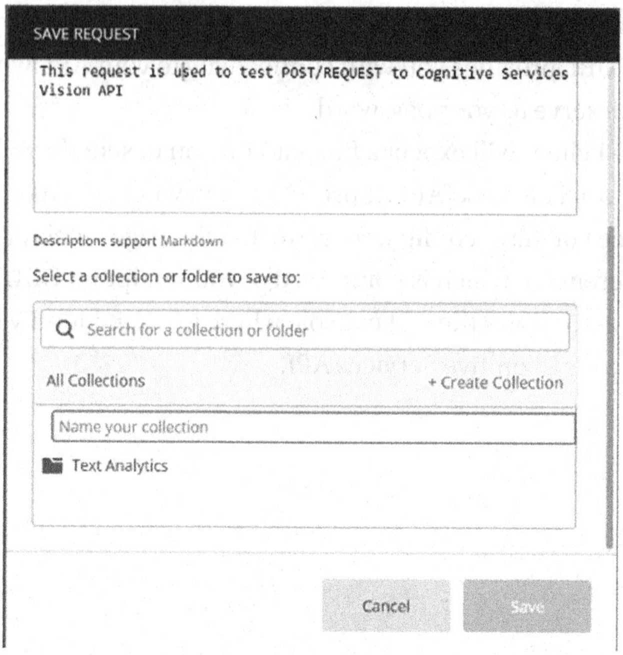

Figure 2-16. *Creating a collection is the first step*

A quick click on the orange check mark to confirm, as shown in Figure 2-17, and you are now able to create requests within your newly created collection.

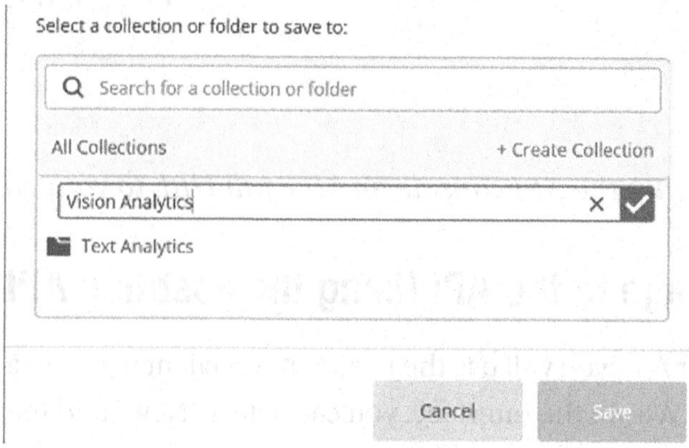

Figure 2-17. *Click to confirm, and you can now create requests*

Once you are on the editor, you can select "POST" from the Request Type dropdown on the left and enter in the full API URL that you previously retrieved from the preceding API console. As shown in Figure 2-18, the full URL for your Cognitive Services is expected in the POST URL field.

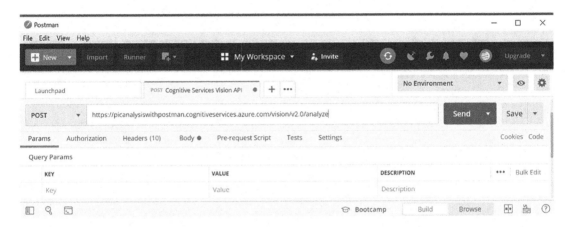

Figure 2-18. *The full URL is expected in the POST URL field*

The values for the headers used by the POST form are key/value pairs. The "Value" field for the key "Ocp-Apim-Subscription-Key" is where we will add the subscription key, as shown in Figure 2-19. On the "Body" tab, you can indicate whether you are sending text or a file.

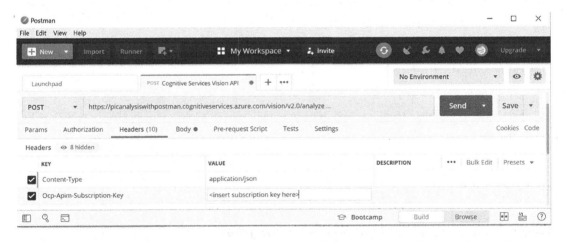

Figure 2-19. *The subscription key is the value for Ocp-Apim-Subscription-Key*

In this instance, you are passing an image file to the Cognitive Services API. The POST request is sending the file "stuffy.jpg," and this file is available in our GitHub repo. As shown in Figure 2-20, the POST request is accepting a URL in the Value field for the key "url."

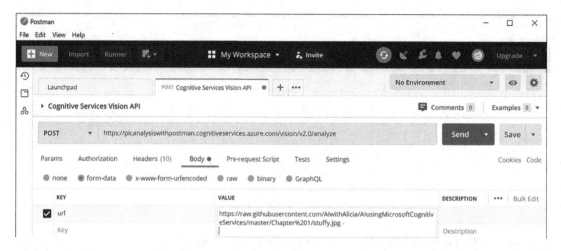

Figure 2-20. *The full URL for the image goes into the Value field*

Once you pass the message by selecting "Send," you will see the interface update as shown in Figure 2-21. The interface will update with the HTTP status message from the Cognitive Services API. You will see a status of "200 OK," response time is noted as 5.53 s, and the size of the message is 513 B.

Figure 2-21. *The JSON response message sent from the API*

As you can see, you can effectively test your Cognitive Services API with very little coding by using the capabilities of the Postman API client.

Docker Containers

Containers are portable and are a virtualization of a server's operating system packaged with the code, runtime, system tools, system libraries, and settings needed to run an application.

A container image is a lightweight, standalone, executable package of software that is portable and allows for applications to run on systems without concern that environmental configurations will be out of sync. Containers have the benefit of allowing programmers to build Cognitive Services models in Azure and reference on a container locally, so that data can be processed within a company's local network and not in the cloud.

Docker is possibly the most common platform on which to build, run, and share containerized applications. Docker Desktop is the best way to get started with Docker on Windows.

If you're unfamiliar with Docker Desktop and want help installing it, then the instructions at the following URL will help:

```
https://docs.docker.com/get-started/overview/
```

When installing Docker, be sure to select "Enable Required Windows Features." This option will turn on Hyper-V and container system requirements. Post installation, you will see "Docker Desktop is Running" in your Windows tray.

Calling the Cognitive Services API on Docker

There are already prebuilt containers that host different types of Cognitive Services. In order to run the Cognitive Services Text Analytics API on a local server, you need to get the image for your machine.

From within a folder on your machine, open a command prompt and enter the following command:

```
docker pull mcr.microsoft.com/azure-cognitive-services/sentiment:latest
```

The Docker image is pulled from Git and downloaded to your local registry. You can now start running the container and services. You can query the Docker image with text sentences and gain your sentiment scores back. The benefits of containerization is that you can run your code and perform machine learning predictions when not connected to Azure.

The docker run command looks as follows. Substitute the data center and API key values and run the following command:

```
docker run --rm -it -p 5000:5000 --memory 4g --cpus 1 mcr.microsoft.com/
azure-cognitive-services/sentiment Eula=accept Billing=https://<datacenter-
here>.api.cognitive.microsoft.com/text/analytics/v2.0 ApiKey=<key>
```

You can interact with the services on the Docker container the same way that you would interact with the services hosted on Azure. Additionally, we can use the same set of steps in the previous section, "Calling the Cognitive Services API with the Postman API Client," to test the API. Following instructions from Figure 2-16, we can make a new Postman request of type POST. We will instead update the endpoint of the service to the following URL:

```
http://localhost:5000/text/analytics/v2.0/sentiment
```

We hope that this chapter will be a reference for you to shortcut your learning and interactions with Cognitive Services.

Summary

The data scientist's toolkit is diverse. Azure Data Studio, SQL Server Management Studio, and Pandas enable programmers to query, cleanse, and manipulate data. Scikit-Learn is a library enabling the construction of machine learning models, and TensorFlow is a library enabling high-performance numerical computation.

Quickstart documentation, including sample code, is available on GitHub for the following programming languages: C#, Python, Go, Java, and JavaScript. You can easily clone repositories with samples and work locally on sample code files.

The Postman API client allows a programmer to perform calls to the Cognitive Services APIs via POST methods and is a helpful tool in allowing us to test connectivity and functionality of Cognitive Services APIs.

The Docker Desktop client enables programmers to create local containers on which they can install Cognitive Services images and link to their Azure services, allowing programmers to perform AI analytics on data and files locally.

CHAPTER 3

Vision

Computer Vision is the ability to extract information from digital images that in aggregate define features within an image. Traditionally, building an image processing model requires the curation of huge amounts of data. For an accurate data model, a representative reference set of images must be collected. Each image must be labeled appropriately, so that the model can be trained. Information is extracted at the pixel level, and pixel details relative to neighboring pixels are considered relevant. Sets of features analyzed with machine learning algorithms enable the identification and location of objects within an image.

Analysis performed at the pixel level runs high on compute requirements. Model building tends to be high on storage and compute requirements. Services and applications that leverage Computer Vision can process and understand visual information as well as trigger events based on features within an image. Applications of Computer Vision involve labeling and classification as well as predictive alerting scenarios.

Many of us have our pictures on social media sites and are aware that capabilities such as auto-tagging our images on photos and unlocking our phones with our faces reveal that someone somewhere is collecting and storing our likeness to build and train models to be able to identify us in images. There are also many fun applications out there that will let us upload an image and provide us with labeling and identification of objects like bugs, animals, leaves, and trees.

When we look at monitoring scenarios in the workplace, we can leverage Computer Vision to recognize more complex situations. For example, hazardous situations in the workplace and noncompliance with safety standards are events previously monitored visually by employees. The power of Cognitive Services enables applications to expedite complex alerting processes by simplifying the process of analysis. Previously, we would have had to rely on a worker to visually recognize that a walkway was blocked or an entryway was obstructed or an object was displaced. Now, these scenarios can be labeled and easily identified by the AI system when they occur.

49

© Alicia Moniz, Matt Gordon, Ida Bergum, Mia Chang, Ginger Grant 2021
A. Moniz et al., *Beginning Azure Cognitive Services*, https://doi.org/10.1007/978-1-4842-7176-6_3

The most commonly utilized components of the Cognitive Services Vision APIs are the Computer Vision and Custom Vision APIs. We will look at these services and some examples of how these can be applied in classification and predictive alerting scenarios. With as little as 20 images, the Custom Vision API will enable the creation of intelligent binary models, where we can determine if an object is or isn't an object of a certain type. We will also have the ability to identify and predict whether objects or features are present via object detection capabilities.

Building on the Vision capabilities, the newly released Face API allows for advanced analysis of features extracted from images of people. This expands our ability to extract data from complex images and applies analytics against complex video and image scenes. Video Indexer also integrates the Vision capabilities and functionality with Text and Voice Analytics. This chapter will showcase the ease of use, simplicity, and power of the Cognitive Services Vision API to embed visual classification and object detection within your applications.

The Vision APIs

The Vision API suite encapsulates functionality for analysis of images and videos that is submitted either via URL or binary format. The results are sent back from the service in the form of a JSON payload.

What Is Included in the Vision APIs

Within the Vision API suite are the Computer Vision, Custom Vision, Face, Form Recognizer and Azure Video Analyzer APIs. A wide variety of results are readily available within the model and require no machine learning programming to access. Table 3-1 gives a description of what each of the APIs provides.

Table 3-1. *Services Available Within the Vision API*

Service Name	Service Description
Computer Vision	The Computer Vision service provides you with access to prebuilt machine learning models that leverage advanced algorithms for image processing.
Custom Vision	The Custom Vision service exposes prebuilt machine learning models with the added benefit of being able to upload your own images for training.
Face	The Face service provides access to prebuilt machine learning models trained. Face attributes are detected, and facial recognition is possible with advanced face algorithms.
Form Recognizer	Form Recognizer enables extraction of typed text from images and stores data in structured key-value pairs, that is, field name and field content.
Azure Video Analyzer (formerly Video Indexer)	Azure Video Analyzer combines capabilities of Vision, Language, and Text APIs and enables translation, image recognition, and object detection.

Reference Architecture

The Vision APIs allow us to send images in binary or via URL to the API endpoint. A variety of characteristics about the objects within each image are returned back in JSON format. With Custom Vision, we can either choose to upload our images for training or process our image locally within a Docker or Kubernetes container. Figure 3-1 shows the interaction between the code you are writing and the Cognitive Services API. Application code sends text or images via a HTTP request to APIs and returns a text JSON response. The Cognitive Services APIs run on infrastructure maintained by Microsoft on Azure, so you do not have to set up servers and install code locally to build and train your own machine learning models.

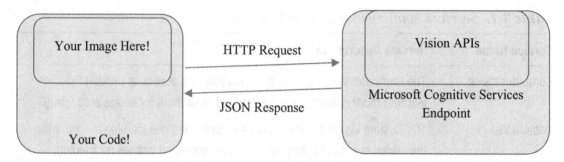

Figure 3-1. Images sent by application code to APIs running on Azure

Computer Vision

The Computer Vision service allows you access to advanced machine learning models that are prebuilt. These models are the result of training and prediction already performed by Microsoft on underlying datasets that are curated and maintained by Microsoft. Not only are the datasets stored by Microsoft but the compute resources to perform the training are also maintained by Microsoft. This benefits a user, because they do not have to collect or create their own digital Intellectual Property (IP) to provide meaningful datasets for relevant machine learning models. Additionally, a user does not have to maintain the hardware/compute needed to process these datasets and generate models. The disadvantage here is these models are already built and the user does not have the ability to tune or influence a model's prediction. As a result, the model results cannot be changed by input from the user.

What types of images are included in the dataset used for Computer Vision machine learning models? A set of images used for building and training a machine learning model is called a domain. Most Computer Vision projects can be created using the everyday objects found in the General domain. The Computer Vision API will return classification and details regarding common objects with fair accuracy, based on varying elements such as how clear the picture is and how unobstructed the object is within the picture. To improve relevance and accuracy of prediction, Microsoft has curated a few more specific datasets, such as the Food, Landmark, and Retail domains. These domains are comprised of images specific to their domains and result in higher accuracy when images include food, landmark, and retail objects, respectively.

Fun Fact   Wikipedia is used as a source to train Cognitive Services models. Do you have a higher accuracy when you pass a Wikipedia image to your Cognitive Services model?

What types of outputs can we expect from the Computer Vision API? The Computer Vision API can be easily leveraged by applications to analyze images and label content, extract printed and written text, recognize brands and landmarks, as well as moderate content. The API is exposed to the programmer, and the application is able to perform recognition using machine learning, without the programmer needing to know any machine learning code!

Custom Vision

For those of us who have more nuanced variations within our images and are looking at basing our machine learning model on specific datasets, Custom Vision will allow the user to upload the training dataset as well as modify prediction thresholds. With the Custom Vision model, we can answer questions such as: How many cows are in the pasture? With the added ability to curate our own dataset, we can upload and label specific images. We can now ask, Which cow is Daisy? Finally, the Custom Vision model allows the user to identify and pass back coordinates of matching objects within the image, allowing us to answer the question, I didn't see Daisy at the feed trough today. Where in the pasture should I look for her?

Face

The Face service provides access to advanced machine learning models that are built specifically on facial attributes, allowing for face detection, person identification, and emotion recognition within images. The service is capable of detecting and counting up to 100 images within an image. Also, a user can curate their own set of images and perform identification and matching of up to one million individuals. Through calculation of coordinates of facial landmarks, such as pupil, eyebrow, eye, nose, lip, and mouth, not only is the service able to calculate hair color, head pose (pitch, roll, yaw), and facial hair. But it can also predict emotions, smile, and gender. Face analysis enables us to answer questions such as: Who was in the picture? What are the demographics of the group within the picture? Was the group happy or sad?

Form Recognizer

Form Recognizer leverages two Vision capabilities, the ability to identify objects and extract coordinates and the ability to extract text from images. Form Recognizer matches text with predefined bounding boxes to automate the digital storage of data within forms. The data is extracted from the form as key-value pairs and table data, and this poses endless possibilities for business workflow automation and paperwork reduction.

Azure Video Analyzer (formerly Video Indexer)

Azure Video Analyzer (formerly Video Indexer) is an Azure Media Services AI solution that leverages Text, Voice, and Vision capabilities to extract information from videos and generate content labels such as keyframes, scene markers, and timestamps for appearance of objects, people, or spoken text and/or phrases. This enhanced meta-tagging enables deep search capabilities across the organization's digital library. These capabilities allow the user to answer questions such as: Which was the last video where this object was present? What was the situation and the context during that video frame? Who was present in the video frame, and what was being said with what emotion?

Computer Vision Service

One of the prominent uses of AI image analysis is the enhancement of metadata associated with media files and the ability to apply this enhancement efficiently and in bulk. Organizations that have been collecting huge amounts of rich media assets need a method to organize, store, and retrieve these assets, as well as to manage digital rights and permissions. The Computer Vision service is useful in automating the process of generating metadata to assist with the classification and labeling of media assets.

Capabilities

The Cognitive Services Computer Vision API offers the following prebuilt functionality derived from machine learning models built by Microsoft and trained on datasets curated by Microsoft:

- *Content Tags* – Identify and tag thousands of common objects within an image based on visual features.

- *Object Detection* – Identify and tag an object within an image and pass back object coordinates for each occurrence.

- *Brand Detection* – Identify and tag thousands of global logos of commercial brands in images or videos.

- *Image Categorization* – Identify and categorize an entire image using pre-defined categories and subcategories.

- *Image Descriptions* – After tagging an image, the results are passed to the Cognitive Services Text API to create sentences representative of what is in an image.

- *Face Detection* – Identify coordinates of a rectangle, gender, and age using the Computer Vision API. Additionally, the Face API provides facial identification and pose detection.

- *Image Type Detection* – Identify whether an image is a line drawing or a clip art.

- *Domain-Specific Content* – Identify additional characteristics within an image that are domain specific.

- *Color Scheme Detection* – Identify whether an image is black and white or color, and extract colors within images.

- *Smart-Cropped Thumbnails* – Identify areas of interest within an image and crop an image and produce optimal thumbnails.

- *Area of Interest Detection* – Identify areas of interest within an image and pass back bounding box coordinates.

- *Printed and Handwritten Text Recognition* – Identify and extract printed and handwritten text and shapes within images.

- *Adult Content Classification* – Identify images with adult, racy, or gory content.

Create a Cognitive Services Resource

Now that you are aware of what can be achieved through the Computer Vision API, you must be wondering, how much work is it to configure and leverage these resources? Say we want to analyze an image and investigate all the possible features that are available to us by default from the already built Computer Vision models. What types of metadata can we obtain from the already trained Computer Vision machine learning models?

My daughter is on her way to collecting 1000 "stuffies"? What types of metadata will I get back from the Computer Vision resource, if I pass the preceding picture to Cognitive Services? Figure 3-2 shows the bounding boxes created by the Vision API. A bounding box identifies the edges of objects within an image.

Figure 3-2. *Example of a bounding box, providing object coordinates*

Let's get started! We can go ahead and create a Computer Vision resource in Azure. After we sign in to Azure Portal via portal.azure.com, we can search for "Computer Vision" via the search bar and select "Computer Vision" from within the Azure Marketplace. Next, we can select a "Location," and as a bonus we still have access to the free tier for Computer Vision services, which will allow us to make 20 calls/minute, up to 5000 calls/month.

Figure 3-3 shows the fields needed to create a Computer Vision resource. After identifying a name, we can select a "Subscription" to assign fees to; a "Location," that is, Azure center within which to host our resources; and the "Pricing tier" for our resources. If we need to create a "Resource group," we can quickly do so via the user interface.

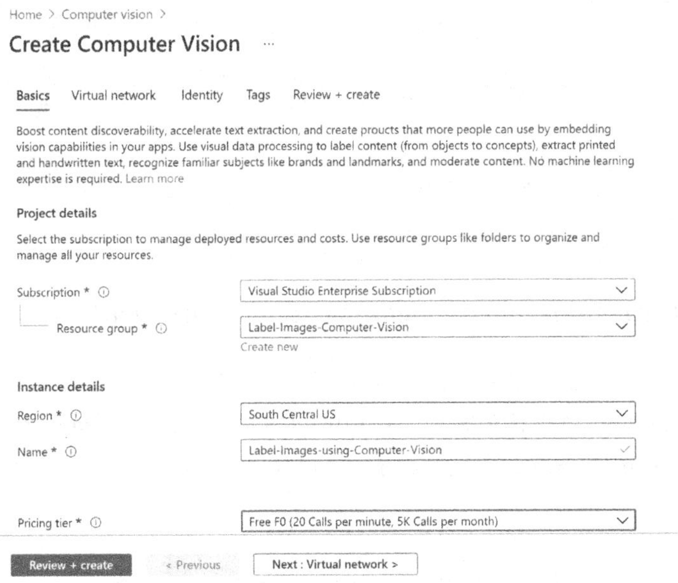

Figure 3-3. *Select Subscription, Location, Pricing tier, and Resource group*

The final step is to review and acknowledge the Responsible AI Notice, customers are responsible for complying with Biometric Data obligations contained within the Online Services Data Privacy Agreement. We can then pass an image over to the Computer Vision API for analysis via a few lines of code in our programming language of choice. We can even test the API directly via the API Reference documentation.

The Computer Vision API page provides an interface for you to test your API. Be prepared to provide your Ocp-Apim-Subscription-Key that authorizes access to your Cognitive Services model. We can get that information from Azure Portal by selecting our resource and navigating to the Keys and Endpoint tab under Resource Maintenance on the left-hand navigation. The first step is to select the region that holds your Cognitive Service. As shown in Figure 3-4, all regions where the service is available are displayed.

Http Method

POST

Select the testing console in the region where you created your resource:

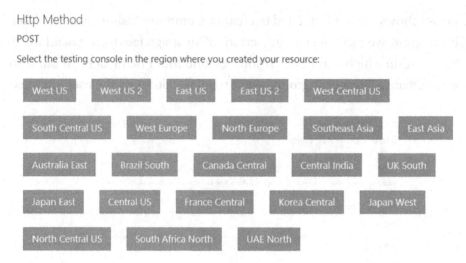

Figure 3-4. Azure regions where Cognitive Services are available

We created our service in the South Central US region, so that's the option we should select. When we click the "South Central US" button, we are redirected to the API console page for South Central US, as shown in Figure 3-5.

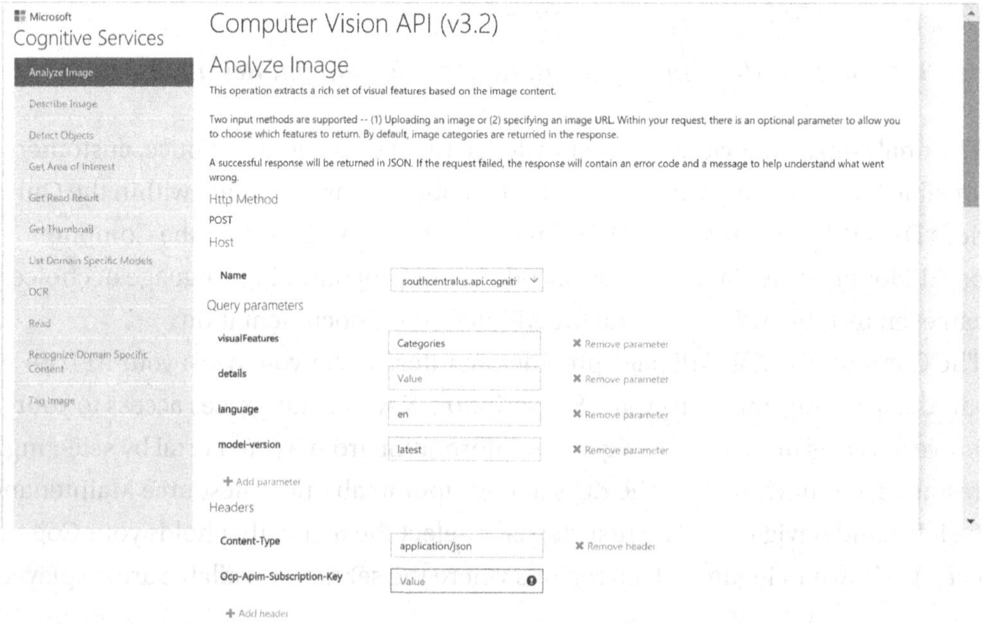

Figure 3-5. The Computer Vision API page provides an interface for testing

The Computer Vision API console page will ask us to provide the parameters that we would otherwise include in an API call. These parameters include

- *Name* – This is prefilled with the URL for the Cognitive Services host for the region, that is, southcentral.api.cognitive.services.

- *Query parameters, visualFeatures* – We can choose to have any of the following features sent back by the API: Adult, Brands, Categories, Color, Description, Faces, ImageType, Objects, and Tags. The default is to return Categories.

- *Query parameters, details* – Default = blank. We can choose whether to use the Landmark or Celebrity domain.

- *Query parameters, language* – Default = "en," for English. We can choose to have results returned in the following languages: "es," Spanish; "ja," Japanese; "pt," Portuguese; and "zh," Simplified Chinese.

- *Headers, Content-Type* – Default = "application/json."

- *Headers, Ocp-Apim-Subscription-Key* – We can find this in the Keys and Endpoint tab under Resource Maintenance.

The last step is to pass over our image file. I have uploaded the image to GitHub, and the file is available via the following URL:

```
https://raw.githubusercontent.com/AIwithAlicia/
AIusingMicrosoftCognitiveServices/master/Chapter%203/1000stuffies.jpg
```

As I fill in my parameters, the POST method is dynamically filled out, and I can see what is sent to the API, as shown in Figure 3-6.

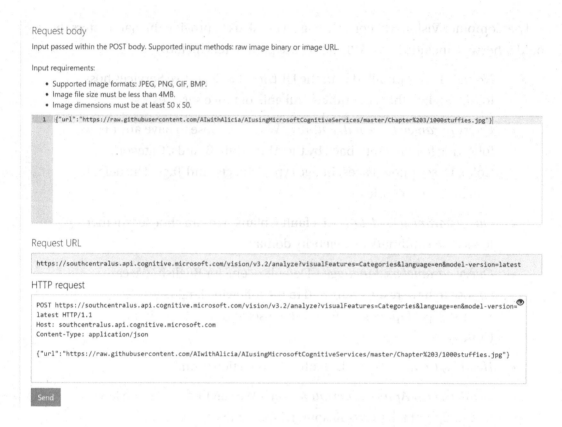

Request body

Input passed within the POST body. Supported input methods: raw image binary or image URL.

Input requirements:

- Supported image formats: JPEG, PNG, GIF, BMP.
- Image file size must be less than 4MB.
- Image dimensions must be at least 50 x 50.

```
1  {"url":"https://raw.githubusercontent.com/AIwithAlicia/AIusingMicrosoftCognitiveServices/master/Chapter%203/1000stuffies.jpg"}
```

Request URL

```
https://southcentralus.api.cognitive.microsoft.com/vision/v3.2/analyze?visualFeatures=Categories&language=en&model-version=latest
```

HTTP request

```
POST https://southcentralus.api.cognitive.microsoft.com/vision/v3.2/analyze?visualFeatures=Categories&language=en&model-version=
latest HTTP/1.1
Host: southcentralus.api.cognitive.microsoft.com
Content-Type: application/json

{"url":"https://raw.githubusercontent.com/AIwithAlicia/AIusingMicrosoftCognitiveServices/master/Chapter%203/1000stuffies.jpg"}
```

Send

Figure 3-6. *The POST method is dynamically filled out using the input URL*

I am now ready to click Send, and the results of my POST request are passed back to me in JSON and when parsed look like Table 3-2. Results returned from the Computer Vision API provide the bounding box coordinates of objects within the image. Categories and confidence levels for labels, adult classification, and colors are also identified.

Table 3-2. *Example of the Results Returned by the API in JSON Format*

Feature Name	Value
Objects	[{ "rectangle": { "x": 2107, "y": 233, "w": 1882, "h": 2366 }, "object": "Teddy bear", "parent": { "object": "Toy", "confidence": 0.849 }, "confidence": 0.841 }, { "rectangle": { "x": 70, "y": 301, "w": 2244, "h": 2637 }, "object": "Teddy bear", "parent": { "object": "Toy", "confidence": 0.859 }, "confidence": 0.852 }]
Tags	[{ "name": "toy", "confidence": 0.9723017 }, { "name": "teddy bear", "confidence": 0.928446651 }, { "name": "rabbit", "confidence": 0.901703 }, { "name": "indoor", "confidence": 0.8606236 }, { "name": "animal", "confidence": 0.8501977 }, { "name": "stuffed toy", "confidence": 0.8379797 }, { "name": "bear", "confidence": 0.7980682 }, { "name": "cartoon", "confidence": 0.7776573 }, { "name": "plush", "confidence": 0.77380836 }, { "name": "stuffed", "confidence": 0.7531785 }, { "name": "rabbits and hares", "confidence": 0.5952502 }, { "name": "cute", "confidence": 0.584520936 }, { "name": "dessert", "confidence": 0.219470531 }, { "name": "shop", "confidence": 0.0927437246 }]
Description	{ "tags": ["indoor", "sitting", "stuffed", "table", "teddy", "bear", "food", "brown", "pile", "holding", "front", "close", "white", "group", "wearing", "large", "woman", "dog", "red", "standing", "kitchen"], "captions": [{ "text": "a group of teddy bears sitting on top of a pile of stuffed animals", "confidence": 0.8870842 }] }
Image format	"Jpeg"
Image dimensions	3024 x 4032
Clip art type	0
Line drawing type	0
Black and white	FALSE
Adult content	FALSE

(*continued*)

Table 3-2. (*continued*)

Feature Name	Value
Adult score	0.003308527
Racy	FALSE
Racy score	0.006383676
Categories	[{ "name": "others_", "score": 0.00390625 }]
Faces	[]
Dominant color background	"Brown"
Dominant color foreground	"Grey"
Accent color	#996732

From these results, we now have additional metadata that allows us to easily label and classify our images. Metadata results also allow us to trigger processes based on the content of images. For example, filtering out and disabling user posts that contain adult content from your public website becomes an easier process when you can trigger automation based on the adult content score of the image.

API Reference: Computer Vision

Since the machine learning models behind Computer Vision are pre-trained, there are many API functions that are exposed and made available to the programmer. Computer Vision v3.2 APIs include the functions listed in Table 3-3.

Table 3-3. *Computer Vision v3.2 API POST and GET Functions*

POST Functions	GET Functions
Analyze Image, Describe Image, Detect Objects Get Area of Interest, Get Thumbnail OCR, Read, Tag Image Recognize Domain Specific Content	Get Read Result List Domain Specific Content

Custom Vision Service

The difference between the Custom Vision API and the Computer Vision API is the ability to influence the outcome of the model by providing your own images for the machine learning model's training.

Capabilities

The Computer Vision API allows us to 1) upload our images, 2) train our model, and then 3) evaluate our predictions.

The Custom Vision Training API provides us with the following methods:

> *Projects* – Create and manage projects, including uploading images and creating and assigning tags.

> *Tagging* – Create metadata labels to "tag" your images with.

> *Training* – Train and test your model using the images you have uploaded, and attach results and configurations to iterations.

> *Export* – Export the Custom Vision project to TensorFlow for Android, Core ML for iOS 11, ONNX for Windows ML, or a Docker container.

The Custom Vision Prediction API provides us with the following methods:

> *ClassifyImage* – Provides classification of an object within an image, based on custom tags and an uploaded dataset.

> *DetectImage* – Provides a bounding box of the detected image.

To add additional value to our insights, there are image "domains" available, which represent models that are trained with specific images:

> *Image classification domains* – General, Food, Landmark, Retail, and Compact domains

> *Object detection domains* – General, Logo, and Compact domains

---

Note   Compact domains are optimized for use on mobile devices!

Best Practices: Images Used for Training

The results of your model will only be as good as the collection of images that you use to build it. The key to building a robust model is variety! When collecting the various images, you want to be sure to have enough representations taken from the various camera angles, lighting scenarios, and backgrounds that will fairly represent the situation that you will be analyzing your images in, in production, for example, if you are evaluating images of a tractor outdoors where the tractor will be exposed to rain and may or may not be covered with dirt in locations. To be relevant to real-life scenarios, you will want to be sure to include real-life images in the dataset that you use to build your model and not just images of brand-new tractors. Additionally, you will want to include images of various types and sizes as well as include individual and grouped sets of tractors.

We also have to keep in mind that to be compatible with and accepted by the API, images need to have the following characteristics:

- .jpg, .png, .bmp, or .gif format.

- No greater than 6 MB in size (4 MB for prediction images).

- No less than 256 pixels on the shortest edge. Any images shorter than this will be automatically scaled up by the Custom Vision service.

Leveraging Custom Vision for Puppy Detection

With the increased usage of email combined with the usage of apps that allow users to transmit images, there is an increased need to auto-classify and label images within the workplace. The Custom Vision API can be useful in identifying and labeling objects within images.

To be compliant with privacy laws, we are going to utilize a dataset of images that include one of the authors' stuffed animals and family pet. Don't worry. The author's 7-year-old daughter has given permission for the use of their puppy's likeness in this publication!

To get started, we will need

- An Azure subscription

- 15 images of our puppy

- 15 images of our stuffed animals

Yes, only 15 images of each classification to build your Custom Vision model!

CustomVision.ai is the starting point for creating a Custom Vision project. You can get there via `https://CustomVision.ai`.

You will need your Azure account, and you can sign in directly using your Azure credentials. You can create your new project by clicking "New Project," as shown in Figure 3-7.

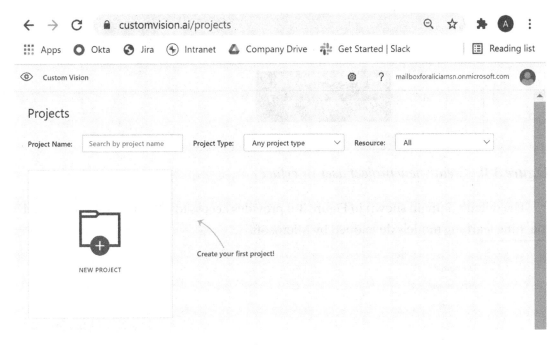

Figure 3-7. *Create a "New Project" on the CustomVision.ai portal*

We are going to create a new project called "Find the Puppy." The project will reference a resource and a resource group that can already be created in our Azure subscription. A resource group is a container to hold a collection of services that operate within the same system. It is recommended that when designing your AI system, you create your services within the same resource group. It is also recommended to create resources within the same region to reduce latency as a result of having to traverse over long distances.

As shown in Figure 3-8, to create a new project, we select from type Classification or Object Detection. We will be creating an "Object Detection" project, in the General domain.

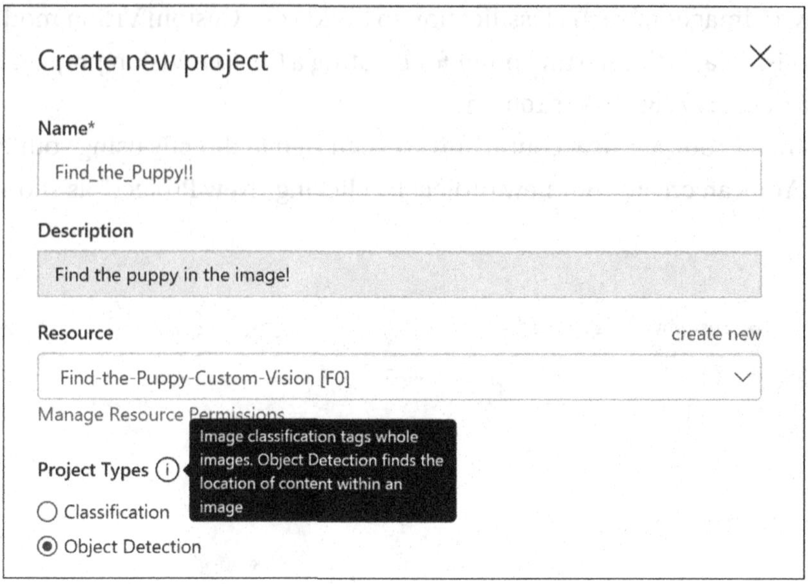

Create new project ✕

Name*

Find_the_Puppy!!

Description

Find the puppy in the image!

Resource   create new

Find-the-Puppy-Custom-Vision [F0]   ⌄

Manage Resource Permissions

Project Types ⓘ

Image classification tags whole images. Object Detection finds the location of content within an image

○ Classification

◉ Object Detection

Figure 3-8. *Create new project user interface*

The default domain shown in Figure 3-9 provides access to APIs driven by advanced machine learning models developed by Microsoft.

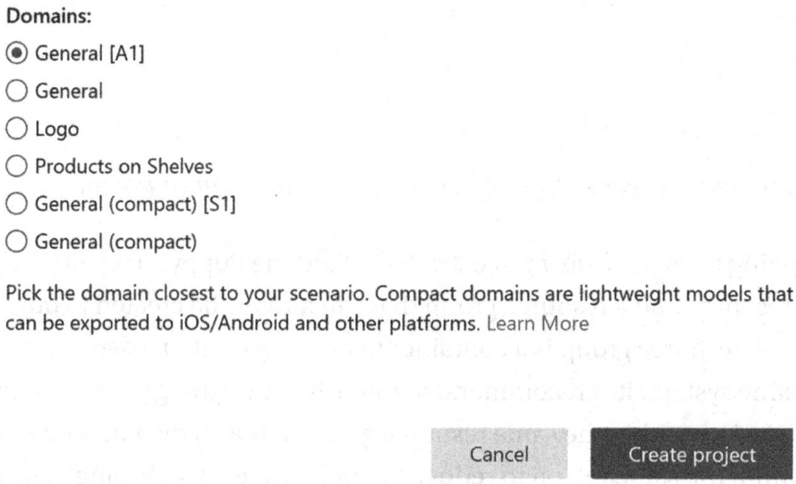

Domains:

◉ General [A1]

○ General

○ Logo

○ Products on Shelves

○ General (compact) [S1]

○ General (compact)

Pick the domain closest to your scenario. Compact domains are lightweight models that can be exported to iOS/Android and other platforms. Learn More

Cancel   Create project

Figure 3-9. *Select a domain from the Create new project interface*

For this example, we will be creating a resource in the South Central US region, since I am based in Houston, Texas, and that is the nearest Azure data center. We can verify configuration details, as shown in Figure 3-10, and select "Create."

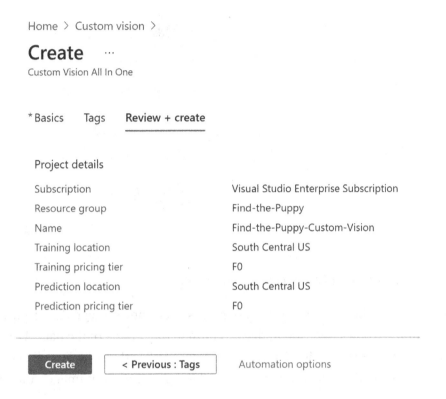

Figure 3-10. *Review parameters before creating the new resource*

Once we have clicked "Create," we will be directed to our project landing page as shown in Figure 3-11.

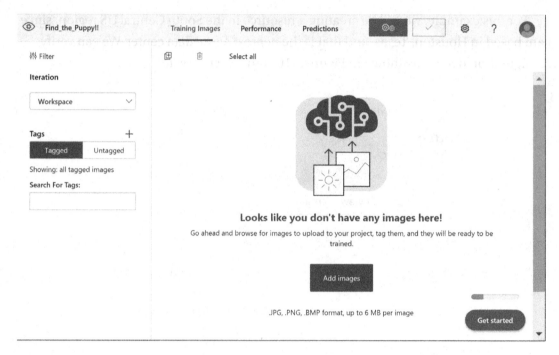

Figure 3-11. *A new empty project on CustomVision.ai*

There are a number of methods to add images to our newly created project. The easiest method is to click the "Add images" button in the middle of the page. We can navigate to the folder containing our training images and Ctrl+A to select all of the images within the folder. The interface will display a preview of our images and prompt for us to upload all images.

After selecting "Add images," we can navigate to our folder containing our images and perform a "Select All." When performing a bulk upload (Figure 3-12), the user interface will preview images within the upload.

Figure 3-12. *Images are previewed when performing a bulk upload*

Custom Vision is a method of creating a machine learning model using supervised learning. This means that we are providing examples, either true or false, of items that are classified as holding certain characteristics. Rather than iterating through a loop of if, then, else statements, we are able to leverage images and apply "tags" as labels. For this exercise, we are able to provide two sets of examples: a group of images representing a collection of stuffed animals and a group of images representing a puppy.

Custom Vision leverages supervised learning. To teach the model the difference between objects, we must label each type of object. Next steps are to create a new tag, so that we can provide an example to the model to learn against. Select the "+" sign to launch the "Create a new tag" pop-up, as shown in Figure 3-13. We first create a tag for "Stuffed Animals" that lets us identify images that are true positives (TP) for the case Stuffed Animals.

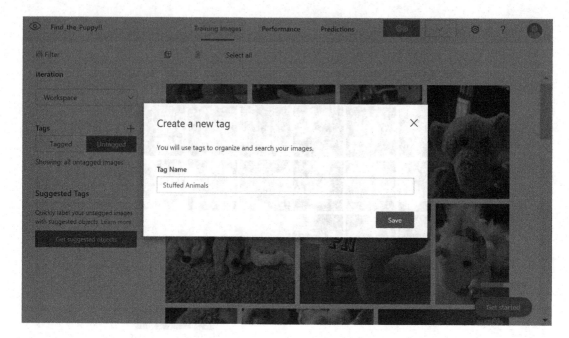

Figure 3-13. *Create a new tag*

Select each image and assign a tag. Figure 3-14 shows both the bounding box and the tag, "Stuffed Animals."

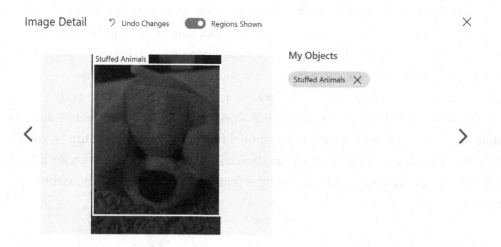

Figure 3-14. *Object within an image labeled with the tag "Stuffed Animals"*

We then step through each of our images and tag appropriately. Figure 3-15 shows all images assigned with the "Stuffed Animals" tag.

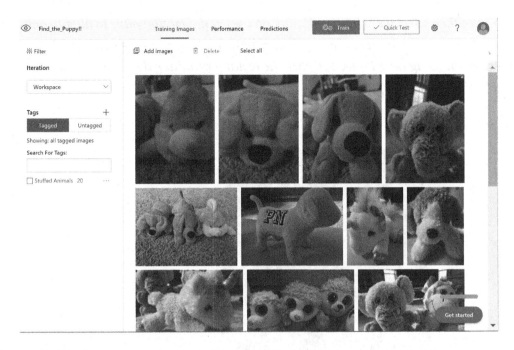

Figure 3-15. *Images can be sorted by assigned tag*

After we have tagged our first batch of stuffed animal images, we can perform the same steps to upload images of our puppy. After uploading, we can create a new tag for tagging puppy pictures, called "Puppy," as shown in Figure 3-16.

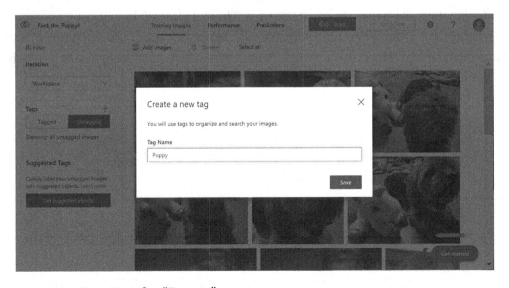

Figure 3-16. *Creating the "Puppy" tag*

We can then select the "Untagged" filter on the left navigation pane to filter to only the new photos that we have uploaded, and we can start labeling images with the "Puppy" tag. As shown in Figure 3-17, we can click the rectangle framing the puppy within the image and select the "Puppy" tag.

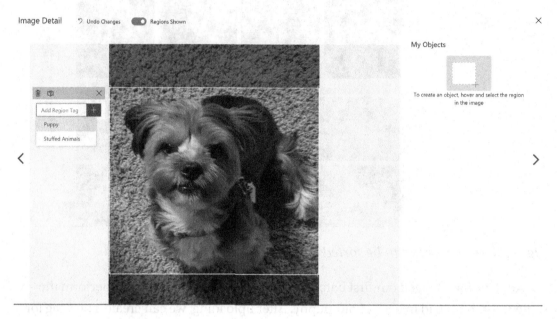

Figure 3-17. *Assign the "Puppy" tag to the bounding box*

After labeling all images, we can perform a "Quick Training" by selecting the green "Train" button on the top of our project. We can choose between "Quick Training" and "Advanced Training." The default "Quick Training" option will allow us to immediately train our model, as shown in Figure 3-18.

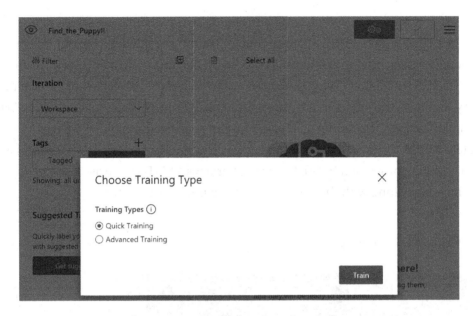

Figure 3-18. *Default to "Quick Training"; select "Train"*

Figure 3-19 shows training results, showing 56.5% precision. We can increase our threshold past the default 30% mark to decrease the occurrence of false positives (FP).

Figure 3-19. *Shows training results, that is, iteration results*

We can now select the "Quick Test" button on the top of the project and either upload an image or pass a URL to our model for training. A quick test of our image correction identifies one stuffed animal with 97.4% accuracy and one puppy in our image with 99.9% accuracy. Why was my accuracy so high? I had a training image that was very similar to the image that was used for prediction. This can lead to issues. With such a small data set, the model will have a harder time predicting new images accurately, due to the lack of diverse images it was trained with. Figure 3-20 shows an image after being processed by the Cognitive Services API. Tags are assigned to objects within the image along with the probability of accuracy for each tag.

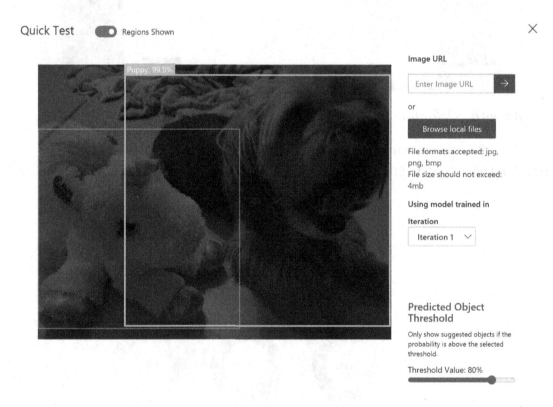

Figure 3-20. *Image post-processing with tags assigned*

Tip   Notice that for the S0 tier, you can upload up to 100,000 images for training and utilize up to 500 tags! That enables you to provide quite a variety of images to your training set!

To programmatically interact with our Custom Vision model, we can access our key and endpoint in the settings, accessible by clicking the gearwheel. As shown in Figure 3-21, model settings are accessible via the gearwheel icon. Easily group the information needed to interact with your Custom Vision model.

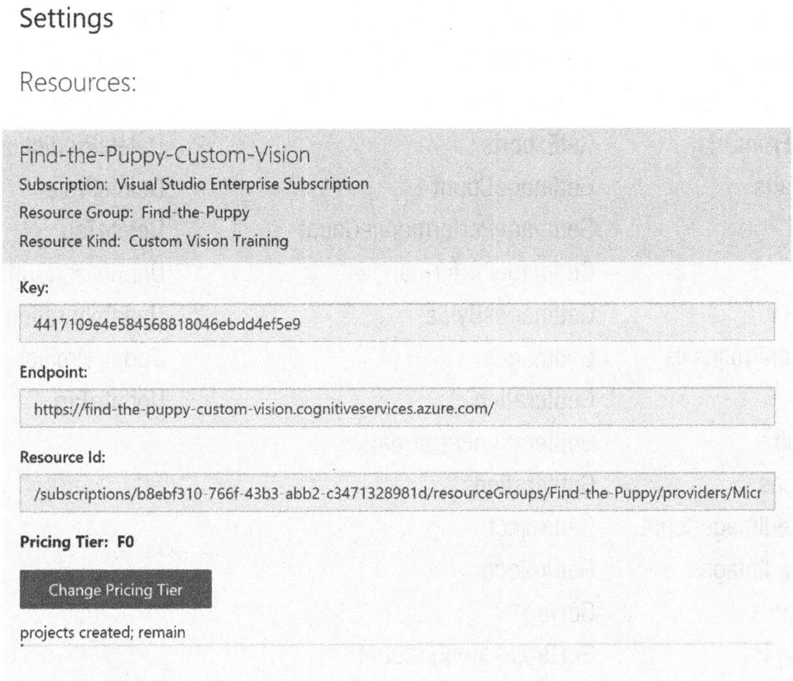

Figure 3-21. *Model settings*

API Reference: Custom Vision Training API

The Custom Vision Training API provides a collection of functions for us to programmatically perform all of the steps that we are capable of performing via the CustomVision.ai portal. With these functions, we are able to upload our images and train and tune our model. Namely, we are able to create, get, update, and delete projects, images, tags, and iterations. The Custom Vision Training v3.4 API includes the functions listed in Table 3-4.

Table 3-4. *Custom Vision Training v3.4 API Functions*

POST Functions	GET Functions	Delete and Patch Functions
CreateImagesFromData	ExportProject	DeleteImageRegions
CreateImageRegions	GetDomain	DeleteImages
CreateImagesFromFiles	GetArtifact	DeleteImageTags
CreateImagesFromPredictions	GetDomains	DeleteIteration
CreateImagesFromUrls	GetExports	DeletePrediction
CreateImageTags	GetImageCount	DeleteProject
CreateProject	GetImagePerformanceCount	DeleteTag
CreateTag	GetImagePerformances	UnpublishIteration
ExportIteration	GetImagesByIds	UpdateIteration
GetImageRegionProposals	GetImages	UpdateProject
ImportProject	GetIteration	UpdateTag
PublishIteration	GetIterationPerformance	
QueryPredictions	GetIterations	
QuerySuggestedImageCount	GetProject	
QuerySuggestedImages	GetProjects	
QuickTestImage	GetTag	
QuickTestImageUrl	GetTaggedImageCount	
SuggestTagsAndRegions	GetTaggedImages	
TrainProject	GetTags	
UpdateImageMetadata	GetUntaggedImageCount	
	GetUntaggedImages	

API Reference: Custom Vision Prediction API

Similarly, the Custom Vision Prediction API provides a collection of functions for us
to programmatically perform all of the steps that we are capable of performing via the
CustomVision.ai portal. We are able to create, get, update, and delete projects, images,
tags, and iterations. The Custom Vision Prediction v3.3 API includes the functions listed
in Table 3-5.

Table 3-5. *Custom Vision Prediction v3.3 API Functions*

POST Functions	
ClassifyImage	DetectImage
ClassifyImageURL	DetectImageUrl
ClassifyImageUrlWithNoStore	DetectImageURLwithNoStore
ClassifyImageWithNoStore	DetectImageWithNoStore

Predictive Weather Monitoring: Image Classification

The Custom Vision API allows us to easily train a machine learning model with as little as 15 images. Can we leverage the API to streamline predictive analytics on real-life scenarios? Is there an opportunity to leverage existing infrastructure including existing cameras and monitoring equipment to gain more readily available insights into advanced weather monitoring?

Tip   Microsoft is active in providing technology, resources, and expertise to empower AI for Good. Interested in learning more about AI for Good? Visit www. microsoft.com/en-us/ai/ai-for-good to learn how Microsoft is supporting AI for Earth, Health, Accessibility, Humanitarian Action, and Cultural Heritage.

Leveraging Custom Vision for Weather Alerting

Traditional numerical weather monitoring involves analysis of extensive datasets. The NOAA's National Operational Model Archive and Distribution System (NOMADS) makes the following NWP and assimilation data available:

> Global Data Assimilation System (GDAS)
>
> Global Ensemble Forecast System (GEFS)
>
> Global Forecast System (GFS)
>
> Climate Forecast System (CFS)
>
> North American Multi-Model Ensemble (NMME)
>
> North American Mesoscale (NAM)

Can you imagine the amount of time, effort, and data involved in building and maintaining a weather system that is dependent on such a huge quantity and variety of data! Figure 3-22 shows the amount of infrastructure that would be needed to store and curate the data needed to build traditional machine learning models for weather data.

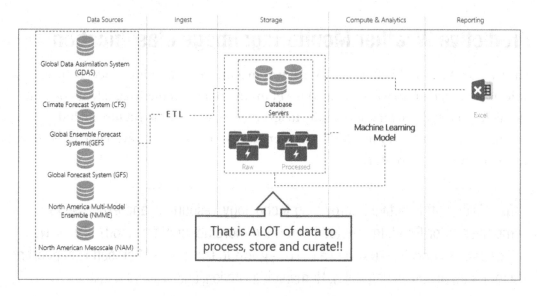

Solution Architecture
Weather Prediction Reporting System

Figure 3-22. Solution Architecture: Traditional Weather Prediction System

Now let's walk through creating a predictive weather model using the Cognitive Services API! We will be walking through an implementation for the solution shown in Figure 3-23. We can navigate to the CustomVision.ai portal and create a new project. We can perform a binary test, to determine whether the channel is flooded or not.

Solution Architecture
Custom Vision - Predictive Weather Alerting

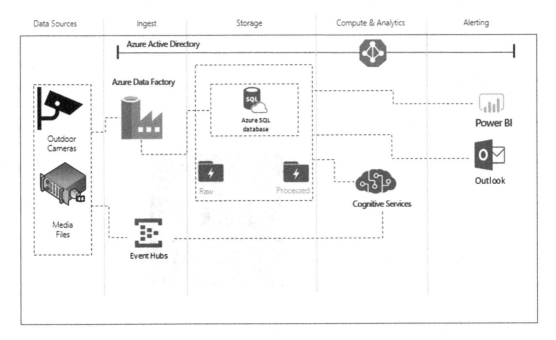

Figure 3-23. *Solution Architecture: Predictive Weather Alerting System*

We start with a set of images taken from a scenario where we consider the channel to be in a "flooded" state. In this scenario, we have collected 20 images satisfying this condition. We will walk through the steps needed to create a new project as shown in Figure 3-24.

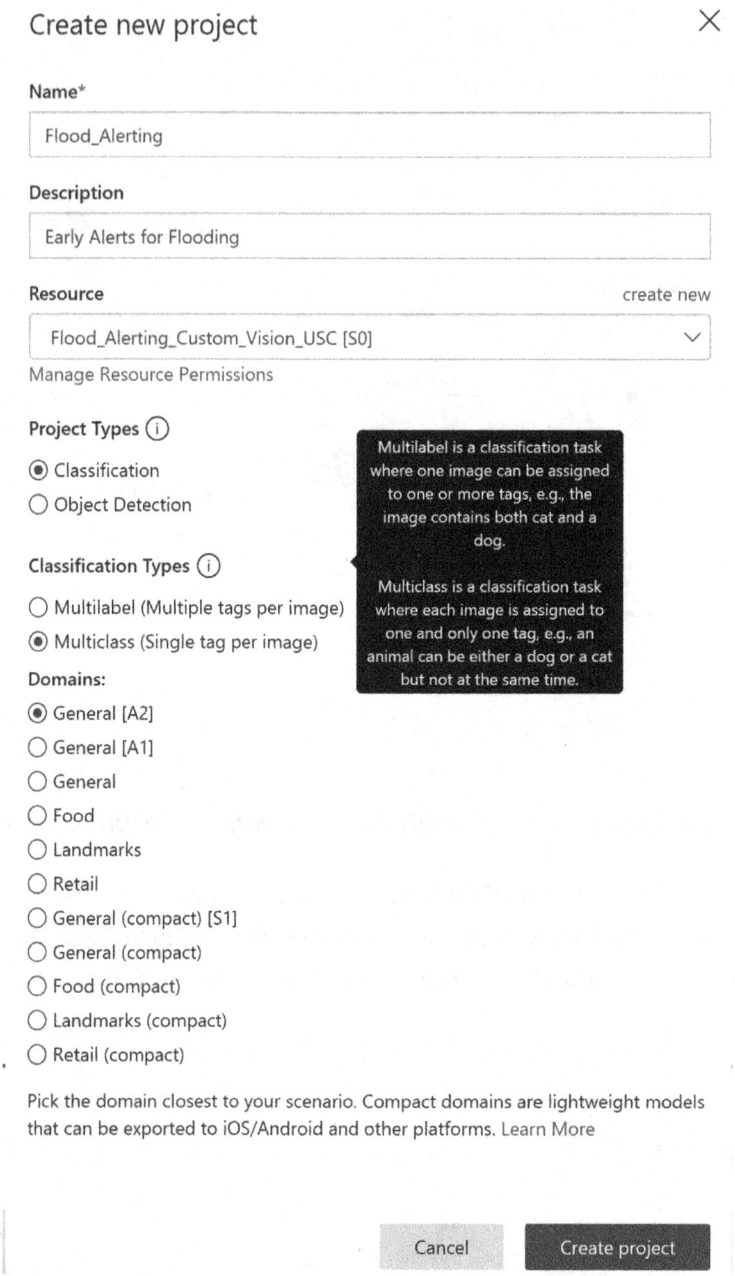

Create new project   ✕

Name*

Flood_Alerting

Description

Early Alerts for Flooding

Resource   create new

Flood_Alerting_Custom_Vision_USC [S0]   ⌄

Manage Resource Permissions

Project Types ⓘ

◉ Classification

◯ Object Detection

Classification Types ⓘ

◯ Multilabel (Multiple tags per image)

◉ Multiclass (Single tag per image)

Domains:

◉ General [A2]

◯ General [A1]

◯ General

◯ Food

◯ Landmarks

◯ Retail

◯ General (compact) [S1]

◯ General (compact)

◯ Food (compact)

◯ Landmarks (compact)

◯ Retail (compact)

Pick the domain closest to your scenario. Compact domains are lightweight models
that can be exported to iOS/Android and other platforms. Learn More

> Multilabel is a classification task where one image can be assigned to one or more tags, e.g., the image contains both cat and a dog.
>
> Multiclass is a classification task where each image is assigned to one and only one tag, e.g., an animal can be either a dog or a cat but not at the same time.

Cancel   Create project

Figure 3-24. *Create a new Custom Vision Classification project*

Click Add images, navigate to the folder holding the images, and perform a Select
All. We can easily "tag" all images at once with the appropriate label. Figure 3-25 shows
"Flooded" example images bulk uploaded successfully.

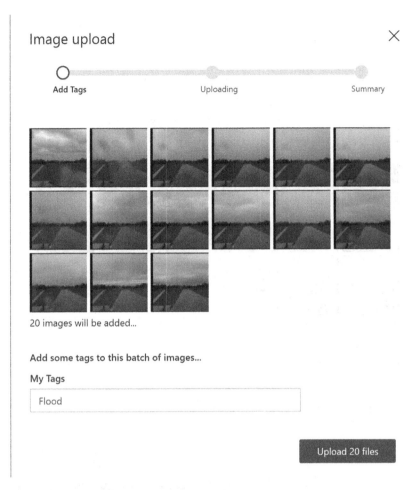

Figure 3-25. *Images of type "Flooded" bulk uploaded and tagged*

Now we can take a set of images taken from a scenario where we consider the channel to not be in a "flooded" state. In this scenario, we have collected 20 images satisfying this condition.

Click Add images, navigate to the folder holding the images, and perform a Select All. This time around, instead of assigning a named tag, we can tag the images with "Negative." This indicates that the images do not satisfy any of the existing tag criteria. Figure 3-26 shows "Negative" example images bulk uploaded successfully.

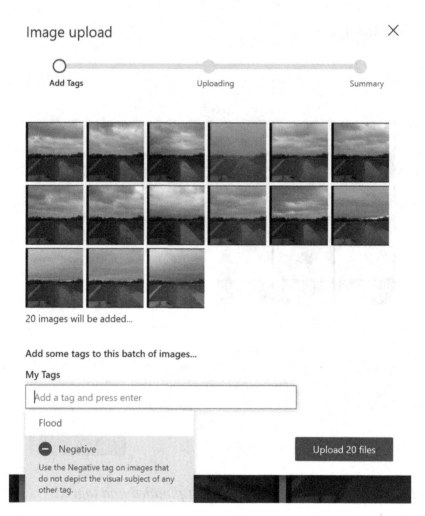

Figure 3-26. *Images of type "Negative" bulk uploaded and tagged*

We now have two sets of images that are available as data sources for our model to learn against. We can select Train and run a quick train.

After the quick train, results are stored with the first iteration. As we make changes to the model, we can save configurations and match them up to their iteration. Figure 3-27 is showing that we have successfully uploaded 20 images labeled "Flooded" and 20 images labeled "Negative." The next step is to click the green gearwheels button to initiate training.

Figure 3-27. *Images of type "Flooded" bulk uploaded and tagged*

At the prompt, we will select "Quick Training," and after a few seconds, we will see the iteration results as shown in Figure 3-28. We have created a machine learning model that takes two sets of images that reflect two different states, "Flood" and "Not Flood." The machine learning model will predict which state an image represents, based on similarities to features within the images used for training. In machine learning, we seek to maximize the true positives and true negatives while minimizing false positives and false negatives (FN). Something is true if it is predicted to be a state and is truly representative of that state. For example, an image of an empty waterway will be a true positive for "Not Flood" and a true negative for "Flood." Similarly, that same image would be a false positive for "Flood" and a false negative for "Not Flood."

On the Performance tab, we can look at the iteration results. Our precision and recall results let us know what percentage our true positive (TP) predictions are, relative to false positives (FP) and false negatives (FN).

The formula for precision is TP/TP+FP, where a model with no false positives will have a precision of 1.0.

The formula for recall is TP/TP+FN, where a model with no false negatives will have a recall of 1.0.

The higher the precision and recall, the better performing our model is. If our precision and recall is consistently low, then we should evaluate whether our cases are clearly defined by our training datasets. Do we have enough clear examples of the waterway in a "Not Flood" state? Or do the example images of "Flood" and "Not Flood" look too alike?

Average precision (AP) is a measure of the model performance that reflects the average performance of the model. What should our accepted minimum values be for precision and recall? A point to consider when building your model is how high your tolerance is for false positives and false negatives. For a medical prediction, I can see how we will want to have extremely high values for precision and recall, while a food classification prediction might not have as many ramifications for incorrectly classifying a pluot as a plum.

Figure 3-28. *Iteration results*

As we look at the results of our training, we notice that we have precision = 100% for "Flood" and recall = 100% for "Not Flood." This reflects that while we may have false negatives, that is, some "Flood" predicted as "Not Flood," we do not have false positives, or some "Not Flood" predicted as "Flood." This may mean that while our images for "Flood" are very similar and distinct for the "Flood" state, the "Not Flood" state may not be as clearly identified.

Now we can test our model. Figure 3-29 is an example of an image representing a "Not Flood" state. Our model has indicated that with 56.9% probability, the waterway is not flooded.

Figure 3-29. *Quick Test with an example of a not flooded image*

Let's also test an image representing a flooded state, as shown in Figure 3-30. We can select the "Quick Test" button and select the path to a local image or pass an image URL. For this example, we will first pass a flood image, and we notice that even though the image is blurred by rain on the camera, we are still able to obtain a result of 74.1% probability that the waterway is flooded.

Figure 3-30. *Quick Test with an example of a flooded image*

Best Practices: Improving Precision

Some things that we can do to improve the accuracy of our predictions are to provide the model with enough valid examples to make a comparison.

Have we provided enough of a variety of images? Is there enough quantity of each variation? Is the lighting consistent with the actual environment, and are the images photographed in similar conditions?

How long did it take us to build and train our model? How much code was involved? How long would it take you to teach a co-worker to perform the preceding steps? The Microsoft Cognitive Services APIs are meant to bring the capabilities of AI to everyone. You can see that by following the preceding steps, it is possible to easily create a machine learning model within hours!

Face

The Facial Recognition API is part of the Microsoft AI Cognitive Services suite. The Face service detects human faces in an image and returns the rectangle coordinates of their locations. Optionally, face detection can extract a series of face-related attributes. Examples are head pose, gender, age, emotion, facial hair, and glasses. The Face API is Azure based and is a subset of the Vision API functionality. To leverage the Face API, an image can be programmatically sent to the API via a number of languages. Along with the image file, the caller can also submit parameters detailing which subset of return values to send back.

Capabilities

- *Face Detection* – Coordinates of human faces within image. Optional face attributes

- *Face Verification* – Evaluates if two faces belong to the same person

- *Find Similar* – Finds faces that either match an image or are similar to an image

- *Face Grouping* – Divides a set of images, so that similar images are grouped

- *Person Identification* – Identifies a detected face against a database of people

Speaker Analysis: Using Face API

How do we increase diversity among our event speakers? To improve something, we must first measure it. The Face API allows us to collect some of these demographics from past events that we may not otherwise be able to collect, which may allow us greater insight into how we can improve these trends.

For this example, we will be using the Face Detection functionality and will explore the option of extracting face-related attributes. For this example, we will request facial attributes for gender and facial hair. The general assumption is that humans classified as "female" with heavy facial hair may possibly be misclassified! The expected response for gender is male or female, and facialHair returns lengths in three facial hair areas: moustache, beard, and sideburns. The length is a number between [0,1], 0 for no facial hair in this area and 1 for long or very thick facial hair in this area.

As an input, I am using the picture shown in Figure 3-31, a photo taken from a recent Houston Area Systems Management User Group (HASMUG) – Azure Edition event that included, from left to right, Ryan Durbin, Billy York, myself, and Jim Reid.

Figure 3-31. *Speaker group, three males and one female*

Once we log into Azure Portal, we can easily create a Face service by selecting it
from the Azure Marketplace. As shown in Figure 3-32, we can then pass an image to
the API. In the following, an example piece of Python code that calls the API in Azure
Notebooks is given. The first two steps of the code assign your API subscription key to
the variable Ocp-Apim-Subscription-Key and define the parameters that are expected as
input by the API.

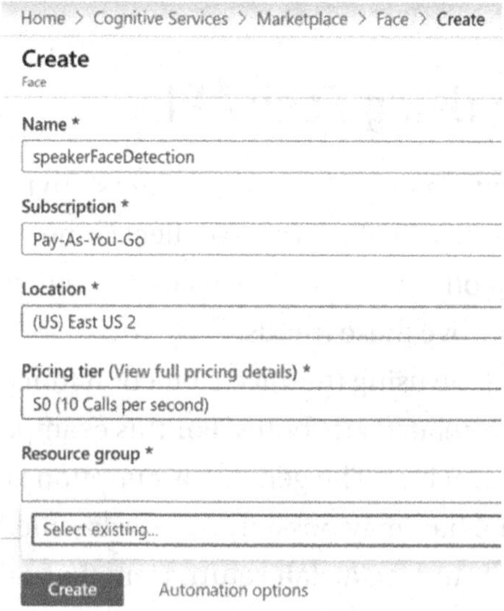

Figure 3-32. *Create a Face API resource*

There are two sections of code needed to programmatically interact with our Cognitive Services API. Figure 3-33 shows the Python pieces of code to open the image file and assign the file contents to a variable. Then finally, the image file contents are passed to the API via a POST request.

```
########### Python 3.6 #############
import http.client, urllib.request, urllib.parse, urllib.error, base64, json, pprint


headers = {
    # Request headers
    'Content-Type': 'application/octet-stream',
    # 'Content-Type': 'application/json',
    'Ocp-Apim-Subscription-Key': '<insert your subscription key here>',
}

params = urllib.parse.urlencode({
    # Request parameters
    'returnFaceId': 'true',
    'returnFaceLandmarks': 'false',
    'returnFaceAttributes': 'gender,facialhair',
    'recognitionModel': 'recognition_01',
    'returnRecognitionModel': 'false',
    'detectionModel': 'detection_01',
})
```

Figure 3-33. *Python code reflecting header and parameter settings*

The Python code shown in Figure 3-34 is an example of the method required to open an image file programmatically and pass contents to the Cognitive Services API. The contents are passed via a POST method.

```
body = open('<insert your filename here>', 'rb').read()
try:
    conn = http.client.HTTPSConnection('<insert your FACE API URL here>')
    conn.request("POST", "/face/v1.0/detect?%s" % params,body,headers)
    response = conn.getresponse()
    data = response.read()
    # pprint.pprint(data)
    prediction_json = json.loads(data)
    pprint.pprint(prediction_json)
    conn.close()
except Exception as e:
    print("[Errno {0}] {1}".format(e.errno, e.strerror))
```

Figure 3-34. *Python code to pass a POST call to the API*

Taking a look at the JSON payload from calling the Facial Recognition API, we can see that there are three males identified and one female. Thankfully, it should be noted in Figure 3-35 that my beard and moustache threshold is 0.0. I was worried that it might not be!

```
[{'faceAttributes': {'facialHair': {'beard': 0.4,
                                    'moustache': 0.1,
                                    'sideburns': 0.1},
                    'gender': 'male'},
 'faceId': '90c19bf8-3203-47b4-a641-1141618b131a',
 'faceRectangle': {'height': 403, 'left': 3158, 'top': 740, 'width': 403}},
 {'faceAttributes': {'facialHair': {'beard': 0.1,
                                    'moustache': 0.1,
                                    'sideburns': 0.1},
                    'gender': 'male'},
 'faceId': '23fcc698-084a-4a3e-8470-45aba7843a73',
 'faceRectangle': {'height': 345, 'left': 600, 'top': 913, 'width': 345}},
 {'faceAttributes': {'facialHair': {'beard': 0.0,
                                    'moustache': 0.0,
                                    'sideburns': 0.0},
                    'gender': 'female'},
 'faceId': 'ae0291bb-aee5-403b-9cdb-0d5bdde6126f',
 'faceRectangle': {'height': 329, 'left': 2241, 'top': 1036, 'width': 329}},
 {'faceAttributes': {'facialHair': {'beard': 0.1,
                                    'moustache': 0.1,
                                    'sideburns': 0.1},
                    'gender': 'male'},
 'faceId': '49c91fb3-f3a6-4ae5-9813-1bc8c5cd50cd',
 'faceRectangle': {'height': 302, 'left': 1581, 'top': 868, 'width': 302}}]
```

Figure 3-35. *JSON payload results, three male speakers and one female speaker*

Optionally, we can also experiment with results when using the Find Similar feature or facial attribute detection. The Face API is another example of how Artificial Intelligence allows us to classify and label data in bulk.

Microsoft's Data Privacy and Security

While the capabilities of the Face API continue to evolve, the caveat of course is, how do we utilize this technology without intruding into the expected privacy of event attendees? Please note: permission has been obtained from each of the individuals in the preceding picture.

As with all of the Cognitive Services resources, developers who use the Face service must be aware of Microsoft's policies on customer data. The Cognitive Services page on the Microsoft Trust Center calls out that Cognitive Services give the programmer control over the storage and deletion of any data stored. Additionally, Face API documentation further details that with facial attributes, no images will be stored. Only the extracted face feature(s) will be stored on a server.

Tip   Ready to get started? Microsoft Learn has an excellent training lab. Identify faces and expressions by using the Computer Vision API in Azure. Available here: `https://docs.microsoft.com/en-us/learn/modules/identify-faces-with-computer-vision/`.

API Reference: Face

Face

The Face API is comprised of methods to Detect, Find Similar, Group, Identify, and Verify. To assist in performing these methods, there are additional APIs that when called provide the containers for facial attributes.

FaceList and LargeFaceList

FaceList allows a user to create a collection of up to 1000 faces within a face list. The list holds the following fields: faceListId, name, userData (optional), and recognitionModel. These lists are referenced by the Face API's Find Similar function. Note: Up to 64 face lists are allowed in one subscription. LargeFaceList allows a user to create a collection of up to 1,000,000 faces.

PersonGroup Person and LargePersonGroup Person

PersonGroup Person allows a user to create a collection of facial features associated with one person of up to 248 faces to perform face identification or verification. The person's extracted facial feature is stored and not the image. LargePersonGroup Person is the container used with LargePersonGroup.

PersonGroup and LargePersonGroup

PersonGroup allows a user to create a collection of up to 1000 person groups, with up to 1,000 persons each using the free tier. The collection holds the following fields: personGroupId, name, userData (optional), and recognitionModel. These groups are referenced by the Face API's Identify function. Note: The S0 tier allows the creation of up to 1,000,000 person groups, with up to 10,000 persons each, and sets of people require the use of the LargePersonGroup.

Snapshot

Allows a user to copy face data to another subscription via backup and restore.

The Face v1.0 API provides access to functions enabling the capabilities for Face, FaceList, Person, PersonGroup, and Snapshot as listed in Table 3-6.

Table 3-6. *Face v1.0 API Functions*

POST Functions	Put and Get Functions	Delete and Patch Functions
Face	**FaceList**	**FaceList**
Detect, Find Similar, Group,	Create, Get	Delete, Delete Face
Identify, Verify	**LargeFaceList**	**LargeFaceList**
FaceList	Create, Get, Get Training Status, List	Delete, Update
Add Face, List, Update	**LargePersonGroup Person**	**LargePersonGroup Person**
LargeFaceList	Get, Get Face, List	Delete, Delete Face, Update,
Add Face, Train	**PersonGroup**	Update Face
LargePersonGroup Person	Create, Get, Get Training Status, List	**PersonGroup**
Add Face, Create	**PersonGroup Person**	Delete, Update
PersonGroup	Get, Get Face	**PersonGroup Person**
Train	**Snapshot**	Delete, Delete Face
PersonGroup Person	Get, Get Operation Status, List	**Snapshot**
Add Face		Delete, Update
Snapshot		
Apply, Take		

Form Recognizer

Form Recognizer extracts information from forms and images into structured data. The service leverages Custom Vision capabilities and extracts text from PDFs and images and stores structured data relative to forms uploaded in the training set. This service enables the automation of data entry and archival of data stored in forms. For example, data from registration or application processes where individuals are asked to fill in the blanks on documents that are formatted to be printable can now be extracted via OCR and stored in databases. Blank versions of the forms are uploaded for training and identifying field labels.

Capabilities

- *Custom Vision* – Identifies and extracts text, keys, tables, and fields. Applies machine learning to generate structured data representative of the original file.

- *Prebuilt Receipt* – Leverages prebuilt models built on images of cash register or bill of sale receipts reflecting sales transactions of goods and/or services. Identifies and extracts data from receipts using optical character recognition (OCR).

- *Layout* – Identifies and extracts text and table structure from documents using optical character recognition (OCR).

Form Recognizer in Action

Please refer to Chapter 6 for an example of Form Recognizer in action!

API Reference: Form Recognizer

The collection of functions available via the Form Recognizer v2.1 API are included in Table 3-7.

Table 3-7. *Form Recognizer v2.1 API Functions*

POST Functions	Put and Get Functions	Delete and Patch Functions
AnalyzeBatch, AnalyzeOperationResult, BusinessCardBatch, BusinessCardBatchResult, ComposeModels, CopyModel, CopyModelAuthorization, CopyOperationResult, CopyAuthorizationResultWithErrors, IDDocumentBatch, IDDocumentBatchResult, InvoiceBatch InvoiceBatchResult, LayoutBatch, LayoutBatchResult, ReceiptsBatch, ReceiptsBatchResult, TrainBatch, TrainBatchWithSubFolders	GetCustomModel, GetModel, GetModels, GetModelsSummary,	DeleteModel

Video Analyzer

Video Analyzer is an Azure Media Services AI solution and leverages AI capabilities across Speech, Language, and Vision. The service allows the programmer to customize and train the models. The service also enables deep search on large quantities of videos and provides organizations the ability to classify and tag media content effectively.

Capabilities

- *People* – Identify specific people within a video and provide analysis about frequency of appearance, video frames, and percentage of presence in the video.

- *Key Topics* – Identify key topics within a video and provide analysis about frequency of appearance, video frames, and percentage of presence in the video.

- *Sentiment* – Identify trend in sentiment throughout the video and provide analysis about frequency of sentiment, video frames, and percentage of sentiment in the video.

- *Key Annotations* – Identify key annotations within a video and provide analysis about frequency of appearance, video frames, and percentage of presence in the video.

- *Transcription* – Embeddable widgets that provide dynamic translation and subtitle services in English, French, German, and Spanish.

Demo the Video Analyzer API

The easiest way to interact with and demo the functionality of the Video Analyzer API is to visit the AI Demos web page maintained by Microsoft. The page provides an interface to interact with the Cognitive Services APIs and can be accessed via `https://aiDemos.microsoft.com`. We can select "Try It Out" underneath the Video Indexer icon and select one of the videos available to see how people, key topics, sentiment, key annotations, and transcription are pulled out of the video.

Media Compliance, Privacy, and Security

While we will be covering AI Ethics in a later chapter, it should be noted that directly on the Video Analyzer page is the following disclaimer: "As an important reminder, you must comply with all applicable laws in your use of Video Analyzer, and you may not use Video Analyzer or any Azure service in a manner that violates the rights of others, or that may be harmful to others. Before processing any videos by Video Analyzer, you must have all the proper rights to use the videos, including, where required by law, all the necessary consents from individuals (if any) in the video/image, for the use, processing, and storage of their data in Video Analyzer and Azure. Some jurisdictions may impose special legal requirements for the collection, online processing, and storage of certain categories of data, such as biometric data. Before using Video Analyzer and Azure for the processing and storage of any data subject to special legal requirements, you must ensure compliance with any such legal requirements that may apply to You." You can visit the Microsoft Trust Center to learn more about compliance, privacy, and security in Azure. Additionally, you can review Microsoft's Privacy Statement, as well as Azure Online Services Terms and the Data Processing Addendum for Microsoft's privacy obligations and data handling and retention practices.

Throughout this chapter, we highlighted how easily we can create Cognitive Services APIs and very quickly interact with the APIs to perform analytics on images. We have walked through recipes for creating Cognitive Services resources via both Azure Portal and the CustomVision.ai portal. We as well walked through scenarios and use cases for object detection.

After reading this chapter, you should be equipped with the knowledge to determine whether to use the Computer Vision API or the Custom Vision API to perform your analysis!

In the next chapter, we will talk about the capabilities of the Cognitive Services Language API!

Summary

In this chapter, you learned

- You want to use the Computer Vision API when you are working with common everyday objects and are performing general classification and metadata management. You want to use the Custom Vision API when you have your own set of images to use for training your data model and you would like the ability to influence the results of your model.

- You can create a Vision API resource within Azure Portal or within the CustomVision.ai portal. You can train and test predictions against your model via the portal, the API documentation pages, or your application code.

- Form Recognizer enables data previously stored as either an image or PDF to be parsed and stored intelligently in databases in key-value pairs.

- You can visit the aidemos.com page to interact with the Cognitive Services APIs without coding or paying for test services.

- There are ethical considerations that influence the architecture and usage of Vision AI capabilities. Microsoft addresses Responsible AI with data privacy and security guidelines, and it is important to consider personal privacy with the storage and use of personal images.

CHAPTER 4

Language in Cognitive Services

Communication is key in business and in life. Despite the challenges posed by COVID-19, our world will continue to be interconnected, and our businesses (and lives) will grow increasingly globalized as time moves forward. While English is commonly understood to be the language of international business, it only makes sense to communicate with people in their native language. Beyond conversation, though, you need the ability to analyze any communication, in a variety of languages.

Polls and surveys may seem somewhat outdated in this digital age and prone to biases. That begs a question – how best to understand a customer or potential customer's feeling about engaging with your brand or product? Tools like sentiment analysis found within the Language API provide unique abilities to get behind the written words and understand the intents and motivations behind those words. The Language API within Azure Cognitive Services opens the door to that ability and so much more. Let's take a look around and see what it can offer us.

Introducing Language APIs

Microsoft states that the Azure Cognitive Services are "sets of machine learning algorithms to solve problems in the field of AI." Interesting, but vague. What specifically, then, do the Language APIs do? Many interesting things! While their very purpose is to unlock a range of possibilities in text-driven communication, Microsoft groups the functionality of the Language API into five key areas: Immersive Reader, Language Understanding, QnA Maker, Text Analytics, and Translator. Table 4-1 describes each of these areas briefly.

© Alicia Moniz, Matt Gordon, Ida Bergum, Mia Chang, Ginger Grant 2021
A. Moniz et al., *Beginning Azure Cognitive Services*, https://doi.org/10.1007/978-1-4842-7176-6_4

Table 4-1. *The Language APIs*

Service Name	Service Description
Immersive Reader	Immersive Reader provides a variety of tools to help users learning to read or learning to read a new language better interact with and understand your content. Beyond that, it enhances the accessibility of your content for people with learning differences.
Language Understanding	Commonly called LUIS (Language Understanding Intelligent Service), this allows you to use the power of natural language processing to bring a conversational experience to apps, bots, and devices with which your users interact.
QnA Maker	QnA Maker allows you to do exactly what it sounds like – create a set of questions and answers over your data. It also adds a conversational element to those questions and answers to offer your users a more natural interaction.
Text Analytics	Text Analytics provides the ability to detect key phrases in the submitted text, identify (and disambiguate) named entities in text, as well as detect positive and negative sentiments in the text that it analyzes.
Translator	Translator provides a language detection service as well as translation from and to more than 60 languages.

As we've emphasized throughout the book, the various APIs within the suite are built to work together. One key area that will highlight that is Chapter 7. That chapter will build on the QnA Maker and LUIS introductions found in this chapter and show you how to integrate those technologies into chatbots. As we've seen with the other APIs that have been discussed so far, this type of integration is made easier by the fact that results are returned from these APIs in the form of a fairly easy-to-read JSON payload.

Immersive Reader

Immersive Reader differs from the other APIs being discussed in this chapter because its sole purpose is to make your content more accessible to readers. While many of the Microsoft-written descriptions of these services are a bit bland, a portion of their description of Immersive Reader really hits home. It states that Immersive Reader helps "emerging readers, language learners, and people with learning differences."

That attitude of service, for lack of a better phrase, is why I chose to open the deep dive portion of this chapter with Immersive Reader. So far, this sounds good, but what does all this actually mean, and what can we do with it? At a high level, you can do things like change the viewable size of the text to adapt it to the reader's vision capabilities as well as display pictures of common words in a "tool tip" format. Drilling down into its functionality, it can do things like highlighting the nouns, verbs, and other parts of speech within the text. For a person learning grammar and sentence structure, this is an important feature to provide. For people learning to read, it can do things like read the content on the screen out loud as well as display the syllables of words to better understand how the words are constructed.

Language Understanding (LUIS)

Language Understanding Intelligent Service fits into a nice little acronym (LUIS), but that acronym is not very descriptive of what the service can actually do. While Chapter 7 will provide a deeper dive into the capabilities of LUIS, I want to familiarize you with the basics here in this chapter so you can dive straight into the great bot-related content later in the book.

Later in this chapter, we'll have a look at a handful of basic examples to help you understand the capabilities of LUIS, but let us set the stage for that with a brief description here. Essentially, LUIS allows you to create custom language models so user interactions with those models are conversational. LUIS breaks down the text into intents ("what do you want me to do?") and entities ("what thing is involved in this request/action?"). These custom models give LUIS the ability to examine all the intents and entities in the text interaction, score those against the model, determine the intents and entities that it is most confident were referred to, and then take any actions determined based on those outcomes. This happens within a natural language conversation, using an application or a bot as its front end, so the gory details of the JSON payload are hidden from the user behind a friendly front end.

QnA Maker

At first glance, QnA Maker can be a bit confusing because a cursory review of the landing page and documentation may make you think that it is very, very similar to LUIS. It, too, is a way to present information to a user interactively in a conversation. While it lacks the flexibility of LUIS, it is far faster to set up and use than LUIS. Chapter 7 will have a section dedicated to QnA Maker, and especially its interaction with chatbots, but this Language API overview is a good chance to review the basics of what QnA Maker can do.

Given the name, it is likely that you have already guessed the service's main function: grouping questions and answers together so, when a question is asked of the service, an appropriate answer is provided if an appropriate answer is available. Those questions and answers are stored in what is referred to as a knowledge base. The knowledge base stores basic sets of questions and answers. Those question and answer sets contained within the knowledge base can be enhanced by "adding alternative phrasing" so the natural language processing picks up on slight variations to the original question. You can also add a follow-up prompt to the answer portion of the knowledge base to kick off more of a multi-turn, conversational experience. That follow-up prompt is returned back to the user via the conversation and prompts the next question.

Now that we have basically defined what the question and answer sets are, that begs the following question: how do the questions and answers get into QnA Maker itself? Once you have created your QnA service within Microsoft Azure, the knowledge base creation process will allow you to populate your knowledge base from a wide variety of sources. The list can change as the service matures and improves, but the generally supported file and source types are URLs (an FAQ page is a particularly suitable source), PDFs, DOC files (Microsoft Word or similar), Excel files (or similarly formatted structured spreadsheets), TXT files, and TSV (tab-separated values) files. You can also manually create question and answer pairs, although that process is more time-consuming than importing a file. As you may imagine, the more structured in a question-and-answer format the file is, the less manual cleanup you will have to do to your knowledge base before deployment. Finally, once your source is imported, you are able to specify the type of "chitchat" your bot will have. Basically, that is your bot's personality. The current choices as of this writing are None, Professional, Friendly, Witty, Caring, and Enthusiastic. The variable chitchat responses are editable if you don't find them to your liking.

Text Analytics

The capabilities of the Text Analytics service may be the most widely known of any of the APIs within Language. Some of that is due to one of its most popular features – sentiment analysis. Put simply, basic sentiment analysis provides a numeric score for the text that it analyzes. In Azure Cognitive Services, the scoring range is between 0 and 1, with a score closer to 1 indicating more positive sentiment and a score closer to 0 indicating more negative sentiment. Version 3 of the API (available in some, but not all, Azure regions

CHAPTER 4   LANGUAGE IN COGNITIVE SERVICES

as of publishing time) provides a different take on this functionality by returning a document label (positive, negative, mixed, and neutral) and a conference score between 0 (low confidence in the assigned label) and 1 (high confidence in the assigned label).

Text Analytics is capable of far more than sentiment analysis of text, however. A recently developed, but quickly maturing, feature is the health feature within Text Analytics. It works to recognize and classify elements like a diagnosis, symptoms, and medication-related information that are present in things like electronic medical records, doctors' notes, and other written artifacts within healthcare. That feature arguably builds on the "key phrase extraction" and "named entity recognition" capabilities that have previously been present within Text Analytics. Named entity recognition attempts to identify entities within the specified text and then classify them into predefined types such as location or person. Key phrase extraction pulls out what it deems to be the most important and most descriptive phrases and terms within the text. Data returned from a key phrase extraction call can also power word clouds and other similar data visualizations.

Finally, Text Analytics provides a language detection ability that can allow you to string together interesting and productive workflows. Language detection analyzes text and returns the languages it believes the document contains along with a confidence score (1.0 being the highest) indicating how high its confidence is in its assessment. It can analyze documents with multiple languages (it will return every language detected along with relative confidence scores for all). It can also analyze text that may be ambiguous (e.g., text containing words that can be found in one or more languages). If you find that language detection is struggling to disambiguate those words, you can add a country hint to your API call. The countryHint parameter accepts a two-letter country/region code and can assist the service's disambiguation efforts by you indicating the country or region that the text/text's creator likely originated. This language detection ability then dovetails nicely with the next (and final) topic on this tour of the Language API: Translator.

Fun Fact   During the 2017–2018 English Premier League season, listeners to the popular Men in Blazers soccer podcast heard a segment entitled "Men in Blazers Mood Table". The standings presented in that segment were calculated using the sentiment analysis portion of the Text Analytics API.

Translator

Translator is well named and does exactly what you think it does. Out of the box, it translates to/from over 60 languages. If you have used the Microsoft Translator mobile app, you have used this API and benefited from the fact that its translation ability is likely far superior to your own! I've certainly benefited from it in my travels.

The latest version (3) of the Translator Text API adds the ability to build customized translation models using Custom Translator for any languages where Translator supports Neural Machine Translation (NMT). NMT is more modern and accurate than SMT (Statistical Machine Translation). These customized translation models can be created via the Custom Translator portal.

The Custom Translator portal gives you a graphical way to upload training data in a wide variety of supported document and file formats and then use those documents to create, train, and test new custom translation systems without writing any code or having deep machine learning experience.

If you prefer to interact with Custom Translator using code, it has its own API as well. Translation is not limited to languages that share a similar or common alphabet, however. Translator also provides transliteration – in layman's terms, converting a word into its counterpart in a different language in a different alphabet. Later in this chapter, we'll take a deeper and more technically detailed dive into some of the capabilities of Translator.

Acquiring and Analyzing Social Media Sentiment

Given how social media has worked its way into nearly every nook and cranny of our society, if you or any organization you're involved with has a social media presence, it is likely a good idea to understand the impression that your social media followers (and others that have mentioned your brand in posts, tweets, etc.) have of you or your organization. Given the speed at which malicious mistakes (or innocuous actions) can be magnified on social media platforms, it makes sense to stay aware of any downturn in social media sentiment so you can respond to negative issues that may have a deleterious impact on you or your business. In this section, we will walk through a basic way to use Azure Logic Apps, Azure SQL Database, and the sentiment analysis ability of Text Analytics to acquire and store tweets and their related sentiment scores. Let's begin!

What Azure Components Are Involved?

We need to create a few Azure resources to work through this example. Let's examine each resource and its role before walking through creating them and using them to view sentiment. The following list details the individual components involved and offers a brief description of what we will do with them:

- *Azure Logic Apps* – Logic Apps are best described as being "event-driven workflow containers." That is not to say they are containers in the Kubernetes or Docker sense; they are more containers in the Tupperware sense. They are plain "boxes" that hold a workflow. Every Logic App is instantiated by a trigger, which is basically the event it is waiting to take place. Our Logic App will search for words in a tweet, then analyze the sentiment of the tweet text itself, and then store that sentiment score and other information in the tweet in an Azure SQL Database for further analysis.

- *Azure SQL Database* – Azure SQL Database is Microsoft's platform-as-a-service (PaaS) database offering. We'll be using a basic deployment of that to store tweet and sentiment data.

- *Text Analytics* – We'll provision a Text Analytics Cognitive Services resource and then connect the logic app to it to automate the sentiment analysis of the text of the tweet.

Azure Logic Apps   Logic apps can be used for much more than just automating Cognitive Services API calls and connecting to social media platforms to ingest textual content. Their uses include connecting to disparate data sources for data movement, connecting to ticketing systems (such as ServiceNow) for ticket and workflow management, rudimentary ETL, and many other uses. They are worth additional research beyond their uses inside of this book.

Create a Text Analytics Resource

In order to perform sentiment analysis on social media posts or other pieces of text, the first component we'll need to provision is a Text Analytics resource. The basics are represented in the following screenshot. In the search box on Azure Portal, if you type "Text Analytics," you will be shown a link to create a Text Analytics resource. Once you click that link, you will see something very similar to the following screenshot.

First, choose a unique name for the resource that is descriptive (and meaningful to you). You're likely to use this resource for many Text Analytics calls throughout your work with the Language API, so it also does not hurt if the name is easy to type as well!

If you have more than one subscription (a work and a personal one, perhaps), you will also need to select the subscription where the resource will reside. It is not required that all of your resources reside in the same subscription, but, in my experience, you'll find everything easier to manage if it is provisioned within the same subscription. You can then manage your different Azure resources using resource groups (which is the bottom field of the five fields in the screenshot). My recommendation would be to use a single resource group per project, but that is certainly an individual choice.

Figure 4-1 shows the initial screen presented to you when you create a Text Analytics resource. The fields included are Name, Location, Subscription, Pricing tier, and Resource group.

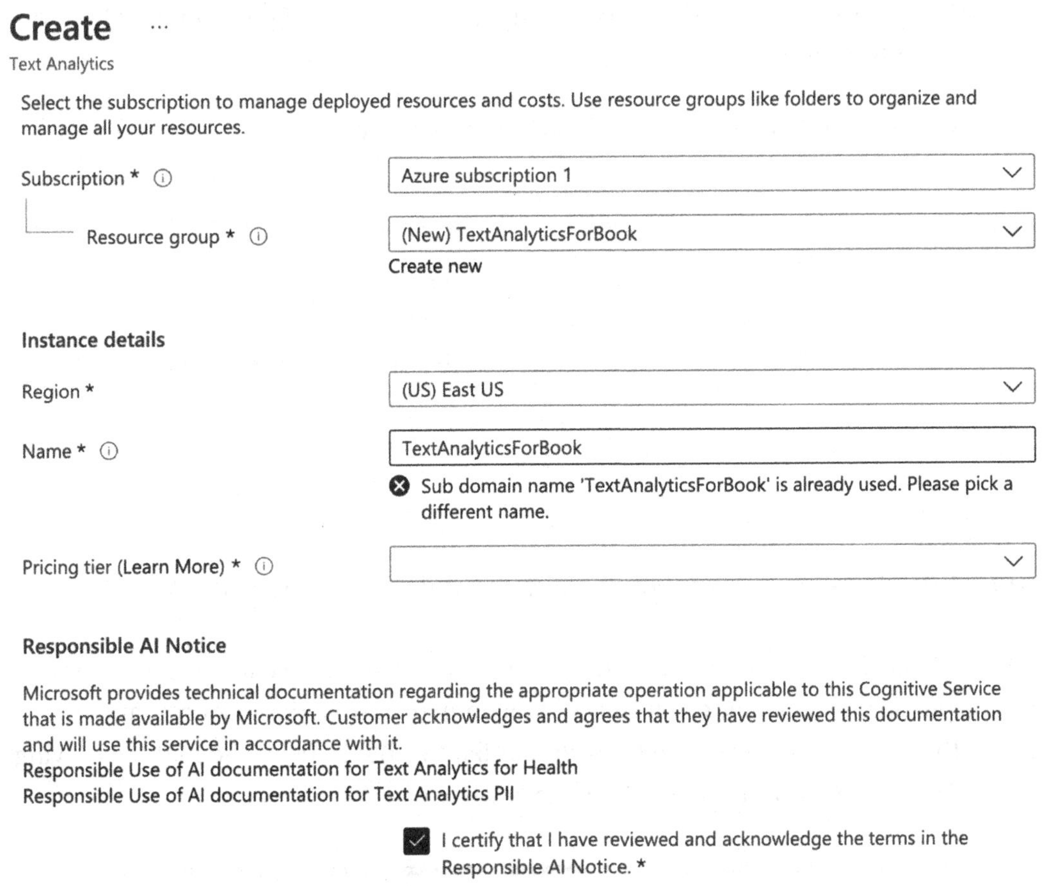

Create · · ·

Text Analytics

Select the subscription to manage deployed resources and costs. Use resource groups like folders to organize and manage all your resources.

Subscription * ⓘ

> Azure subscription 1 ⌄

└─── Resource group * ⓘ

> (New) TextAnalyticsForBook ⌄

Create new

Instance details

Region *

> (US) East US ⌄

Name * ⓘ

> TextAnalyticsForBook

❌ Sub domain name 'TextAnalyticsForBook' is already used. Please pick a different name.

Pricing tier (**Learn More**) * ⓘ

> ⌄

Responsible AI Notice

Microsoft provides technical documentation regarding the appropriate operation applicable to this Cognitive Service that is made available by Microsoft. Customer acknowledges and agrees that they have reviewed this documentation and will use this service in accordance with it.
Responsible Use of AI documentation for Text Analytics for Health
Responsible Use of AI documentation for Text Analytics PII

☑ I certify that I have reviewed and acknowledge the terms in the Responsible AI Notice. *

Figure 4-1. *An example of the first step in creating a Text Analytics resource*

The final two fields to discuss are Location (i.e., the Azure data center where you want your resource to reside) and Pricing tier. In general, you want to select the Azure closest to your typical physical location to decrease latency, but nearly all of the regions will give you roughly equivalent performance. The pricing tier is a more critical choice, but the choice depends on the nature of the project on which you are working. For a personal project, the free tier may be sufficient. If not, the first standard tier is likely sufficient. The tiers above that are typically geared toward more professional projects, but that is not to say a hobbyist could not select those as well. Kudos to you if the scope of your personal project demands one of the higher pricing tiers, and hopefully this book is a great benefit to your ambitious goals!

Once you have made all your selections, you will be presented with a Create button. Once you click that, you should eventually see the following screenshot indicating successful creation of the Text Analytics resource.

Figure 4-2 shows the Overview screen displayed at the end of the deployment of your Text Analytics resource.

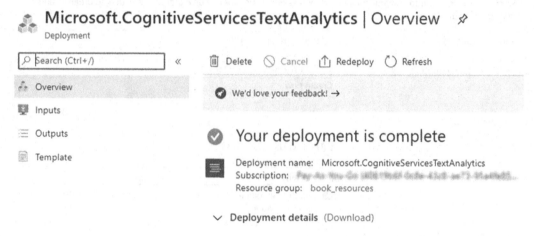

Figure 4-2. *Overview screen showing a successful Text Analytics deployment*

Just to quickly confirm the creation of your resource, use the search box in Azure Portal, type the resource name you selected for your Text Analytics resource, and confirm that it displays in the search results similar to what is displayed in the following. Figure 4-3 shows you how to search for your newly created resource.

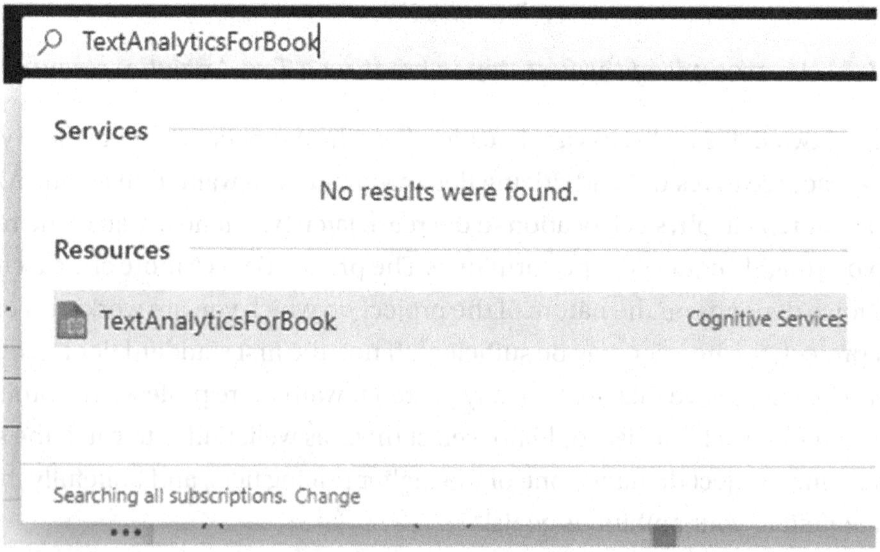

Figure 4-3. *A successful search for your new resource*

Last but not least, select your resource in Azure Portal, look to the leftmost blade in the portal, and click Keys and Endpoint under the Resource Management portion of the blade. Copy and paste both keys into a secure location – these will be essential to have for any subsequent connection to your resource. Figure 4-4 shows an example of viewing your API keys (without revealing mine!).

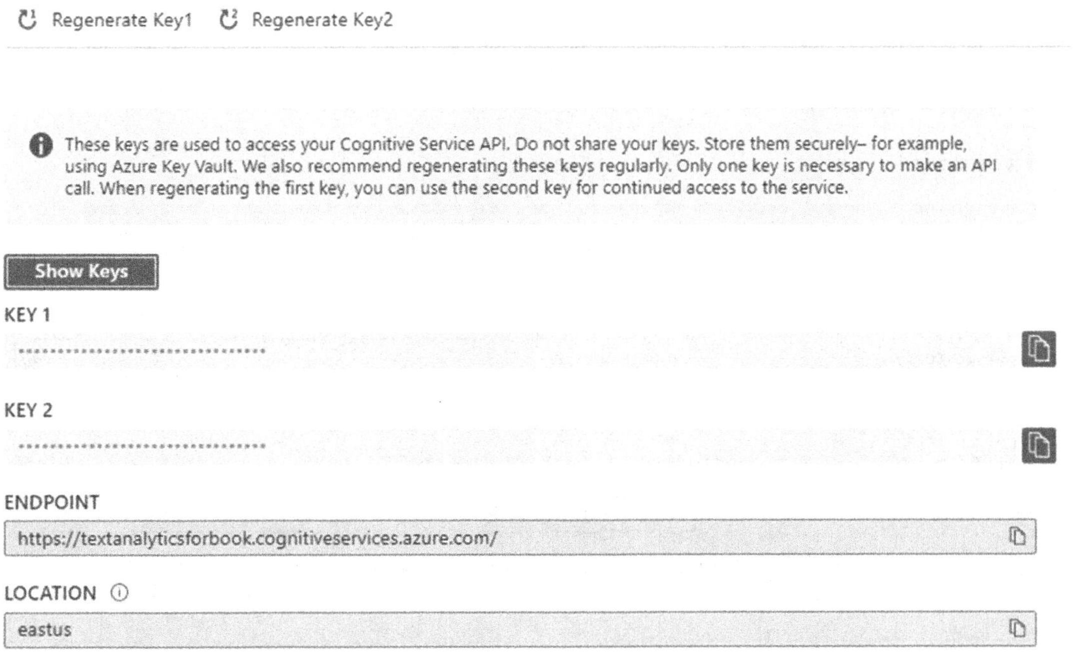

⟳ Regenerate Key1   ⟳ Regenerate Key2

ⓘ These keys are used to access your Cognitive Service API. Do not share your keys. Store them securely– for example, using Azure Key Vault. We also recommend regenerating these keys regularly. Only one key is necessary to make an API call. When regenerating the first key, you can use the second key for continued access to the service.

Show Keys

KEY 1

....

KEY 2

....

ENDPOINT

https://textanalyticsforbook.cognitiveservices.azure.com/

LOCATION ⓘ

eastus

Figure 4-4. *Screen showing how to view API keys, location, and endpoint*

Create an Azure Logic App

Next, let us create the Azure Logic App that will facilitate calls to the Text Analytics resource. After clicking the Create a resource link found on the top left of Azure Portal, you will be taken to the Azure Marketplace window. In the Search Marketplace area, type Logic App and click it. That will bring you to a screen with a Create button. Click that button, and you will see a screenshot like the next image in this chapter, Figure 4-5.

While the layout is different than the last resource creation screen, you will make fairly similar choices. You will first select your subscription, and then, after that selection, the Resource group dropdown is populated with all the resources in your subscription.

After making those choices, select a name for your logic app. Remember, relevant and descriptive names help as you increase the number of Azure resources in your subscription. You will typically select Region in the "Select the location" area and then again select the Azure data center (Location) closest to your typical physical location. Figure 4-5 shows the Basics screen you encounter when first creating your Azure Logic App.

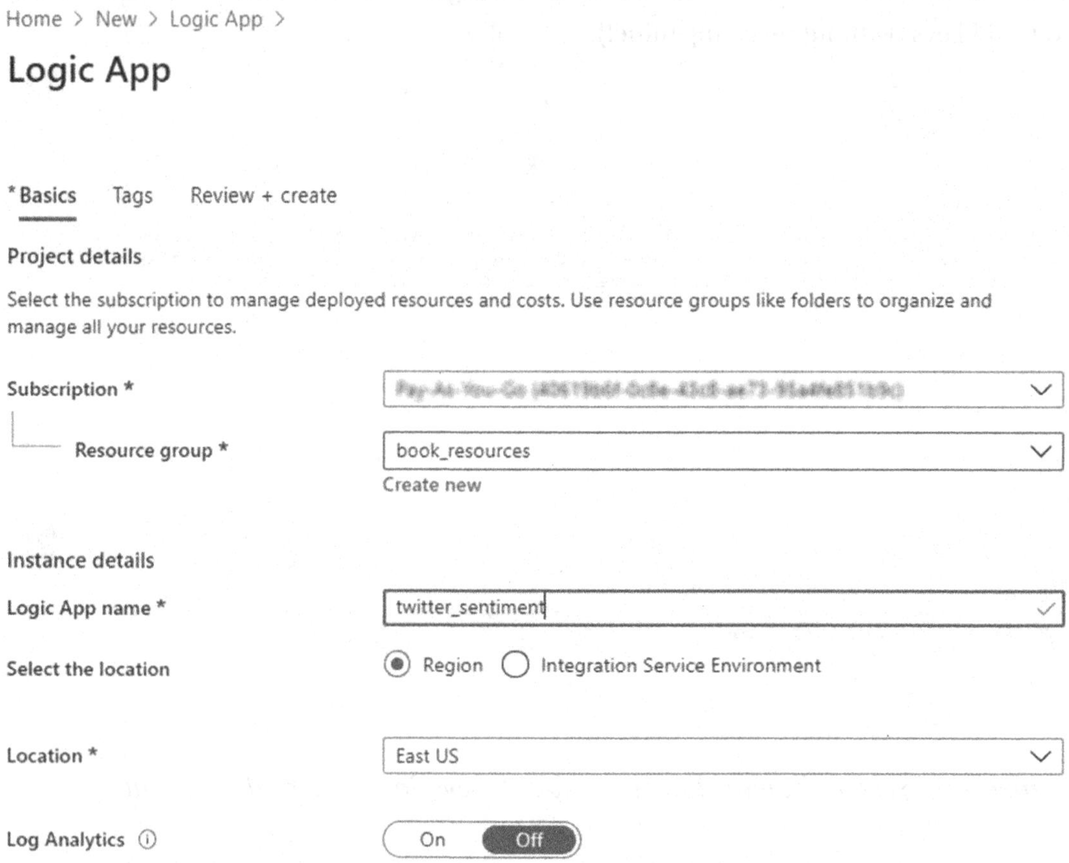

Home > New > Logic App >

Logic App

*Basics   Tags   Review + create

Project details

Select the subscription to manage deployed resources and costs. Use resource groups like folders to organize and manage all your resources.

| Subscription * | Pay-As-You-Go (806 19b6f-0c6e-43c8-ae73-95aff485 1b9c) ⌄ |

| └─ Resource group * | book_resources ⌄ |
Create new

Instance details

| Logic App name * | twitter_sentiment ✓ |

Select the location   ⦿ Region   ◯ Integration Service Environment

| Location * | East US ⌄ |

Log Analytics ⓘ   (On   **Off**)

Figure 4-5. *Specifying the basic details for your Azure Logic App*

Finally, you can select whether or not to turn on Log Analytics. For one-off projects (or projects in an environment where you're not using Log Analytics to do event monitoring and gather event logs), leave that selection off. If you are using Log Analytics and want events and notifications from this logic app to be a part of that, move that selection to On.

Next, we will create the logic app itself. As logic apps are event-driven workflows presented as software-as-a-service (SaaS), the first choice you make is the event that drives the workflow. In this case, select the "When a new tweet is posted" option when you enter the logic app creation screen. Figure 4-6 shows us an example of some common triggers for Azure Logic Apps.

Start with a common trigger
Pick from one of the most commonly used triggers, then orchestrate any number of actions using the rich collection of connectors
When a message is received in a Service Bus queue
When a HTTP request is received
When a new tweet is posted

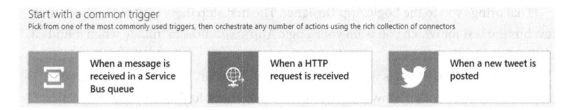

***Figure 4-6.** Commonly suggested triggers for Azure Logic Apps*

After selecting Twitter, you will need to confirm that the logic app is permitted to connect to Twitter and authorize it to do so (in a screen similar to the one shown in Figure 4-7).

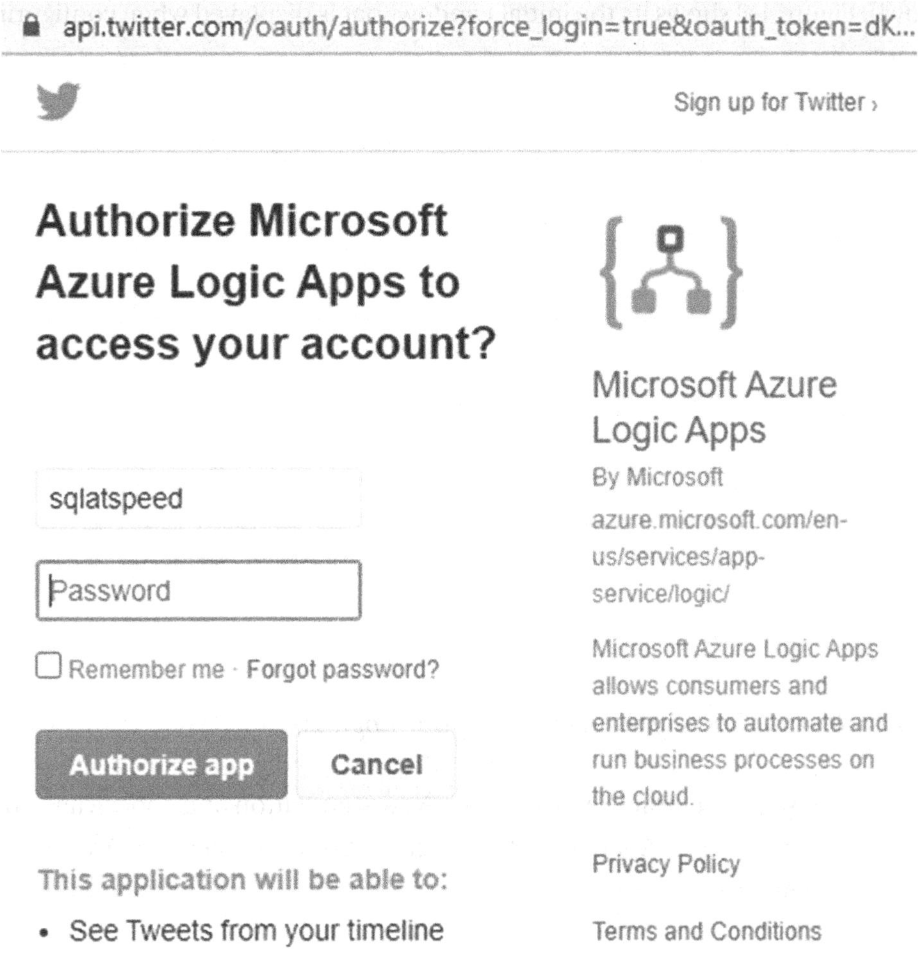

api.twitter.com/oauth/authorize?force_login=true&oauth_token=dK...
Sign up for Twitter ›
Authorize Microsoft Azure Logic Apps to access your account?
sqlatspeed
Password
Remember me · Forgot password?
Authorize app
Cancel
This application will be able to:
• See Tweets from your timeline
Microsoft Azure Logic Apps
By Microsoft
azure.microsoft.com/en-us/services/app-service/logic/
Microsoft Azure Logic Apps allows consumers and enterprises to automate and run business processes on the cloud.
Privacy Policy
Terms and Conditions

***Figure 4-7.** Authorizing an Azure Logic App to access a Twitter account*

That brings you to the Logic App Designer. The first step there is to put in the Search text box the text for which you want your Logic App's workflow to initiate when it finds it via the Twitter search API.

The next option, which is worth a moment of discussion, is the time interval you are asked to select under "How often do you want to check for items?" That interval is less fixed than the text would have you believe. If your logic app is set to search for a common search term, it will continue instantiating instances of your logic app as long as the search API returns results. The timer you are setting on this screen, in a case like that, will never reset, and you will just see constant runs of your logic app. That timer only resets when there are no active runs of your logic app. That concept is very important to understand when you are trying to manage and contain your Azure costs (ask me how I know!). Figure 4-8 shows us the initial window that is displayed when configuring this trigger.

Figure 4-8. *Configuring the trigger for your Logic App*

Once that step is complete, you click the New step button and choose an action. In this case, you type Text Analytics into the search box and select Text Analytics as depicted in Figure 4-9.

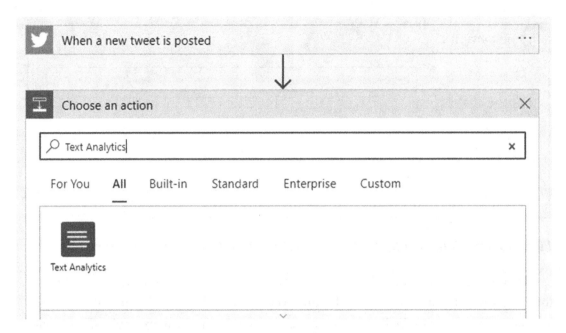

Figure 4-9. *Adding Text Analytics to your Logic App*

After you click that, you will select the Detect Sentiment action from the list that appears. If it does not appear in the first few selections, type the first few letters to rearrange the list and present you the choice you need to select. Once you have done that, you are presented a screen to connect your Logic App to the Cognitive Services resource provisioned earlier in the chapter. You provide a connection name of your choice, the account key (Key 1 as discussed and shown earlier in the chapter), and the site URL. The site URL is the endpoint that can be found on the Overview blade of the Cognitive Services resource provisioned earlier. Figure 4-10 shows the screen used to specify the location and details of the Text Analytics resource you wish to use.

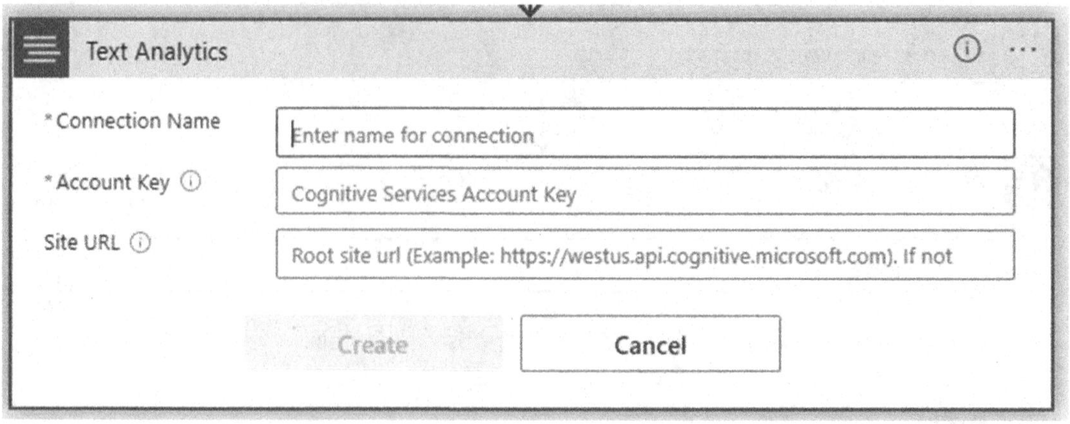

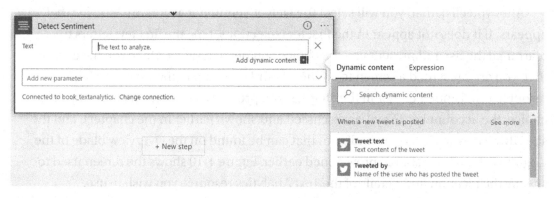

Figure 4-10. *Configuring the Text Analytics action to connect to the new resource*

That selection will then bring up a prompt for the text you would like to send to the Detect Sentiment action. You will want to click the Add dynamic content hyperlink and select Tweet text in the list that appears. Figure 4-11 shows an example of how that would look.

Figure 4-11. *Displaying how to send the tweet text to the Detect Sentiment action*

Following that, click the New step button once again, type SQL Database into the search box, and select the Insert row action. The naming of this step can be a bit confusing because, as you hopefully noted earlier, we are planning to store our sentiment data in an Azure SQL Database. Because you can connect a logic app to both a traditional SQL Server and an Azure SQL Database offering, however, these actions are named as SQL Server. Figure 4-12 shows how this slightly confusing screen looks at the time of publishing.

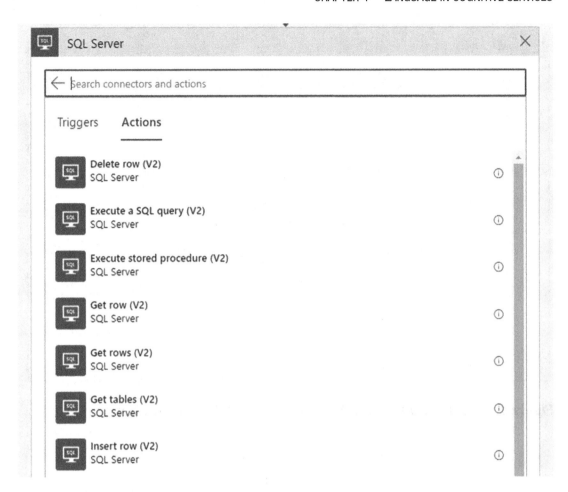

Figure 4-12. *Selecting a database-related action for your Logic App*

A deep dive into the nuances of the different types of authentication to SQL Server instances and Azure SQL Database deployments is outside the scope of this book, but I believe it is helpful to display the screenshot shown in Figure 4-13 and the choices available so that you can record that information from the flavor of SQL Server or Azure SQL that you have deployed or to which you have access.

After authenticating to the server specified in the previous step seen in the following, make sure you select a table. In this case, we will be using the table we can create via the .sql script in our GitHub repository. Figure 4-14 shows how to select the table you will be using to store your information.

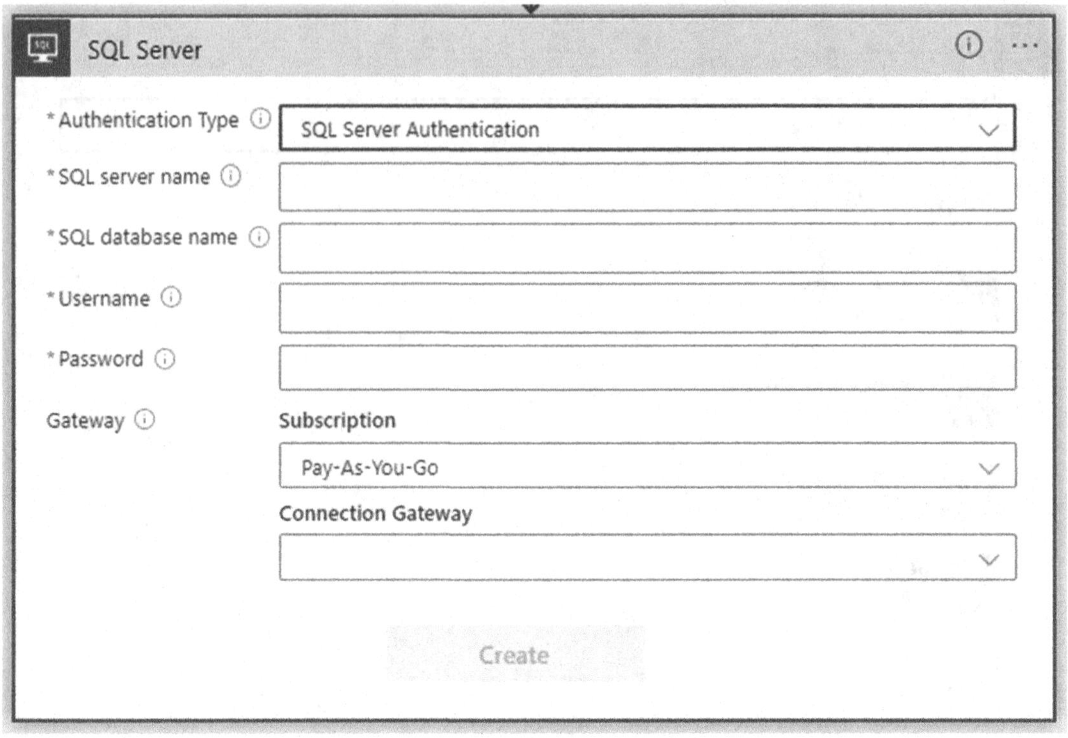

Figure 4-13. *An example of a SQL Server/Azure SQL authentication screen*

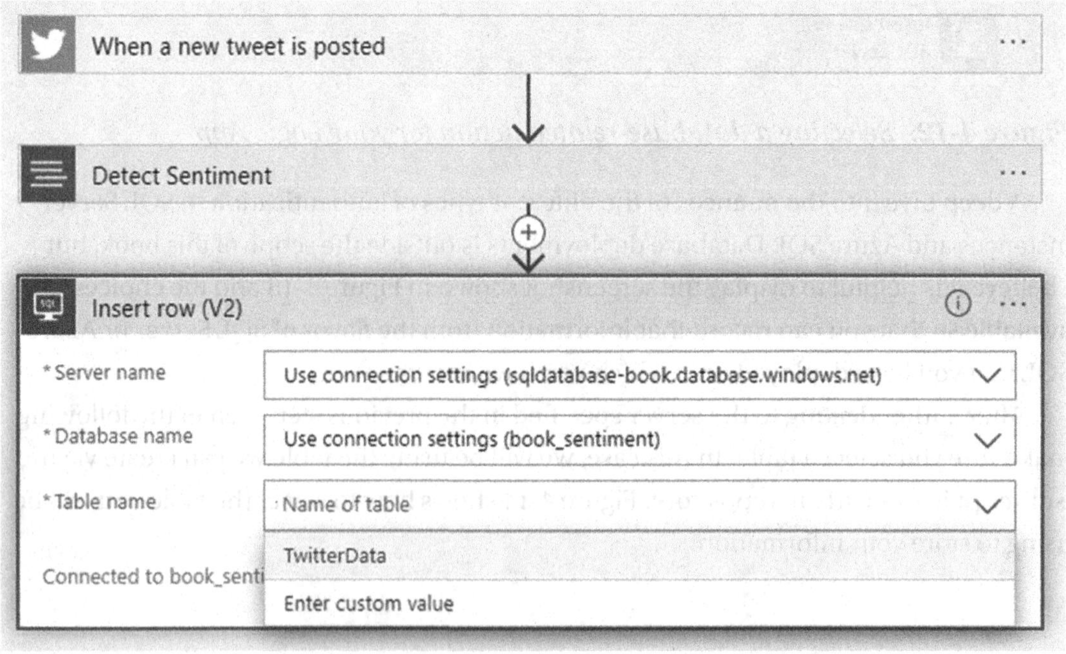

Figure 4-14. *Connecting to a database and choosing the table to use for storage*

That should prompt you to select the columns to populate. Select every column. Figure 4-15 shows what this selection screen should display.

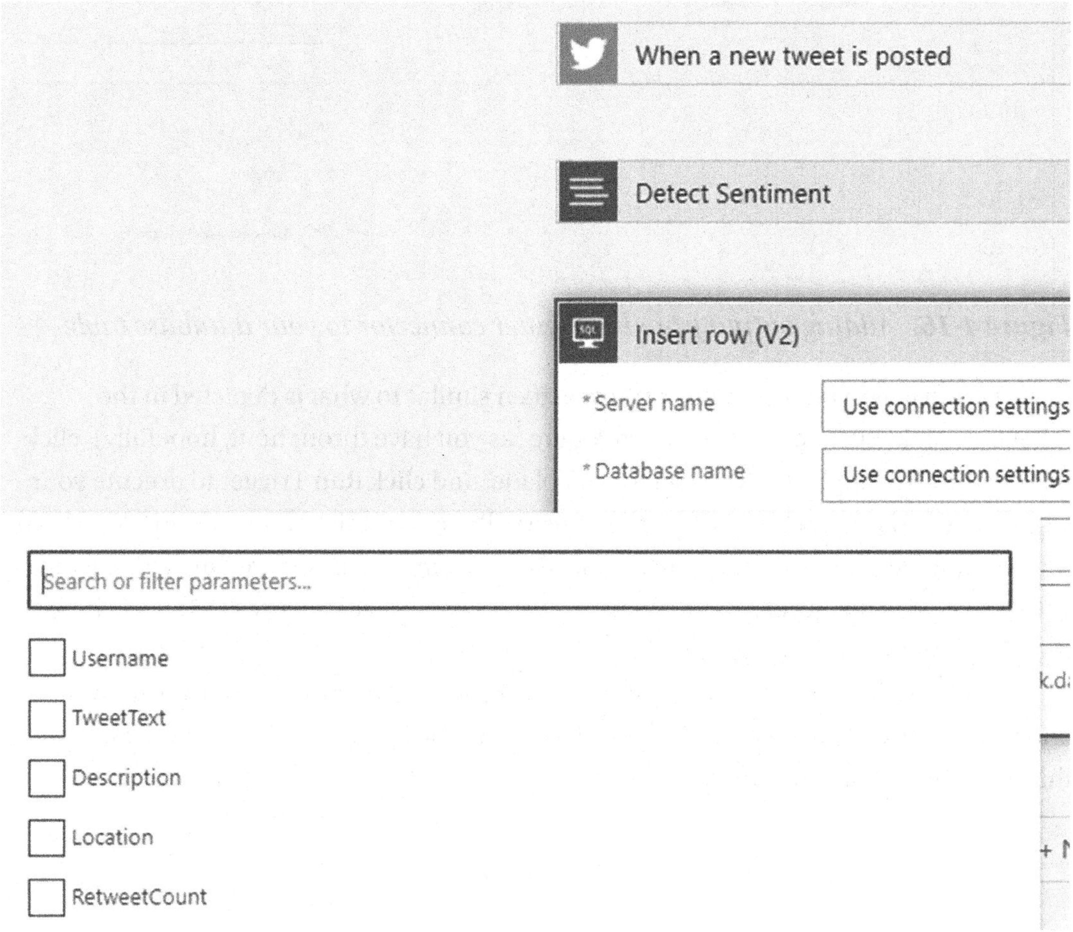

When a new tweet is posted

Detect Sentiment

Insert row (V2)

* Server name   Use connection settings

* Database name   Use connection settings

Search or filter parameters...

Username

TweetText

Description

Location

RetweetCount

Figure 4-15. *Screen used to select table columns for use in database actions*

You will then see a list of the chosen columns. As you place your cursor in the field next to the column name, you will see the Dynamic content pop-up explored earlier. Select the matching fields from Dynamic content (typing a few letters of the field if you don't immediately see it pop up). To that end, typing "score" will give you the score output from the Detect Sentiment action even though it may not immediately pop up in the list. Figure 4-16 shows example an example of how to add dynamic content to the database action.

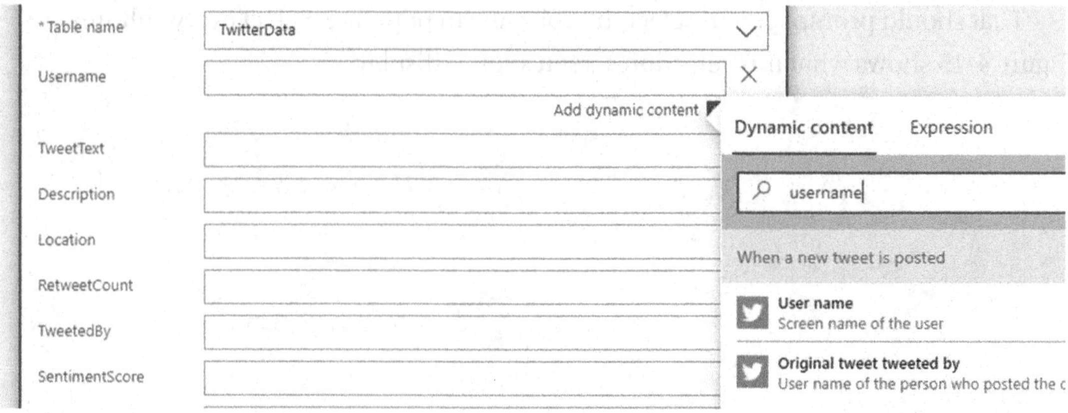

Figure 4-16. *Adding output from the Twitter connector to your database table*

That will leave you with an Insert row screen similar to what is depicted in the following. Make sure you save your work here (as you have throughout, hopefully), click back to the logic app's Overview on the left blade, and click Run Trigger to execute your trigger. If it finds a search term, your database will now contain all the information about the tweet(s) found. Ensure that you disable the logic app once you are finished playing with it so it does not accidentally run up unnecessary charges. Figure 4-17 shows how the completed screen should appear.

Congratulations! You are successfully analyzing tweets for their basic information and sentiment score as well. As discussed earlier in the book, a multitude of uses await this newfound ability of yours.

Figure 4-17. *Dynamic content from Text Analytics and Twitter configured to be stored in your database table*

Driving Customer/User Interaction Using Azure Functions and the Language API

At this point in our explanation, we know how to create a logic app and call Cognitive Services resources with the logic app. We also know how to store the sentiment score and other useful information of the social media source to which we are connected. Let us build on our new knowledge of sentiment analysis but leverage another Azure resource to trigger an alert to us if a tweet with a sentiment score below a certain value is found. We are going to introduce Azure Functions as one way to do that and work in a bit of code-based development as well.

As a sidenote, this section's walk-through assumes the creation of a logic app containing a "When a new tweet is posted" trigger and a Detect Sentiment action. We will then plug in the Azure Function at that point in the workflow. Go ahead and create the two steps of that logic app, save it, and return to this walk-through. To make this walk-through easier to see in action, select a negative term for the search term in the logic app trigger.

Azure Function Used by an Azure Logic App

Similar to other resources created earlier, we can begin by clicking Create a resource in Azure Portal. After typing a few letters of function, we will see Function App. Click that icon. Figure 4-18 shows an example of how this should look.

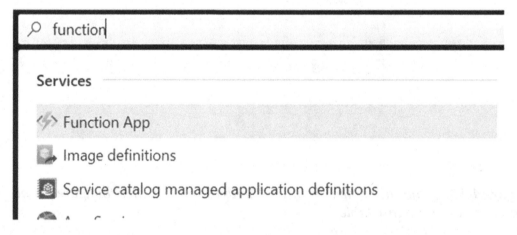

Figure 4-18. *The first step to creating an Azure Function*

Clicking Function App will bring up the screen depicted in Figure 4-19. This may look similar to creating a logic app or other resources, and in many ways it is. We will still select our subscription and resource group, name our Function App, and select a region. As this app simulates sending an email to a customer service resource asking them to contact a customer, I've named it customercontact. You are welcome to choose your own name as well. What is new here, though, are the Publish, Runtime stack, and Version fields. While there are several choices here that detailed Azure Function learning resources explain in depth, to continue building our skills using this walk-through, we are going to select the Code radio button in the following, use .NET Core as our runtime stack (as we intend to use C# as our language of choice within the Function App), and use the latest version (3.1 at the time of publishing) of our runtime stack. Figure 4-19 depicts these choices on the Project Details screen.

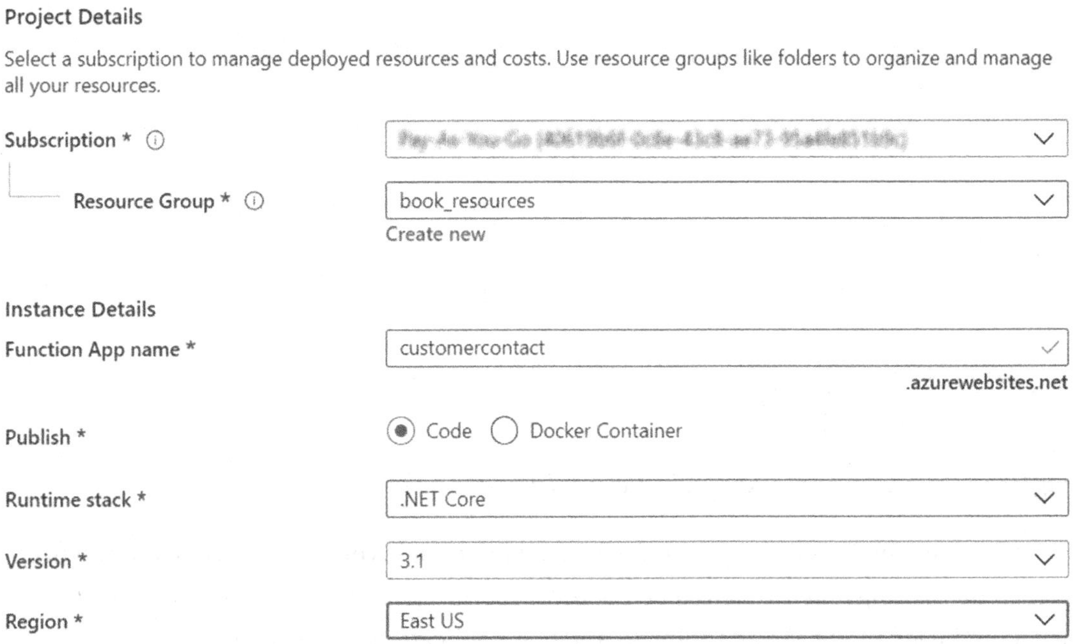

Project Details

Select a subscription to manage deployed resources and costs. Use resource groups like folders to organize and manage all your resources.

Subscription * ⓘ		∨
Resource Group * ⓘ	book_resources	∨
	Create new	

Instance Details

Function App name *	customercontact	✓
	.azurewebsites.net	
Publish *	⦿ Code ◯ Docker Container	
Runtime stack *	.NET Core	∨
Version *	3.1	∨
Region *	East US	∨

Figure 4-19. *Specifying the details of our Function App*

After creating our Function App, we will be prompted to create a function as depicted in the next screenshot. I've simply named it LogicAppTrigger and left the authorization level defaulted to Function. Figure 4-20 shows an example of how this should look.

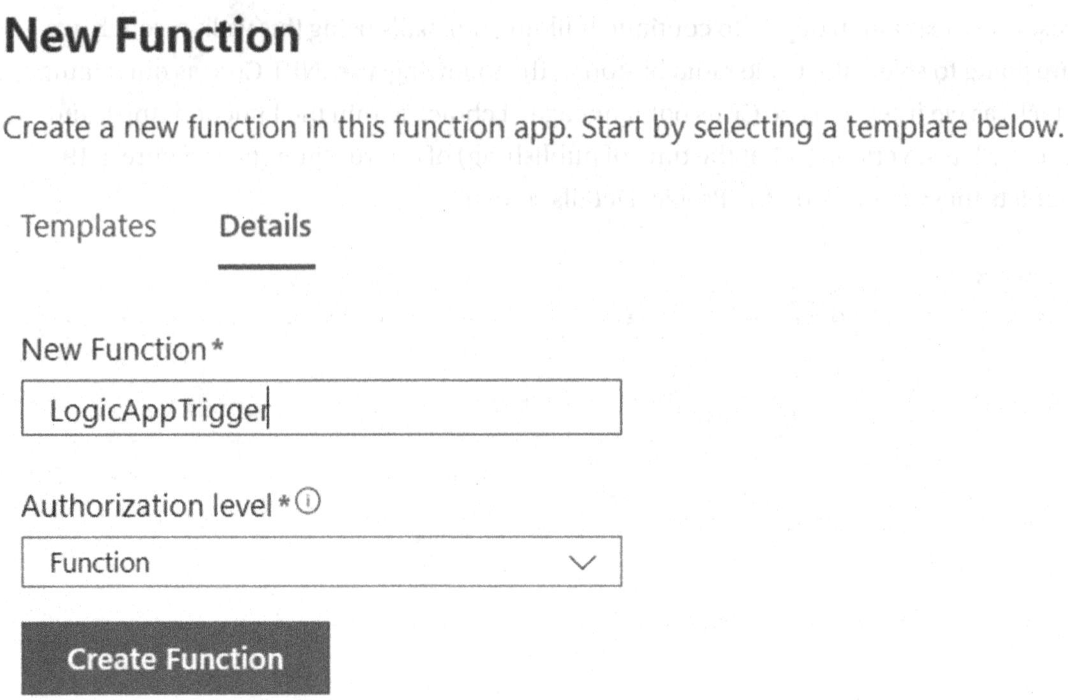

New Function

Create a new function in this function app. Start by selecting a template below.

Templates   **Details**

New Function *

LogicAppTrigger|

Authorization level * ⓘ

Function   ⌄

Create Function

Figure 4-20. *Creating the new function inside the Function App*

After the function is created, you will be taken to the Overview pane of your function. Search for "settings" in the search box of the Overview pane if you are not able to see it on the left blade. We are going to add two application settings here by clicking the New application setting button: APPINSIGHTS_INSTRUMENTATIONKEY and APPLICATIONINSIGHTS_CONNECTION_STRING. For those settings, we will once again need the endpoint/URL of our Text Analytics resource as well as the API key we used in our original logic app. The URL will be placed in the connection string app setting, and the key will be placed in the instrumentation key app setting as depicted in Figures 4-21 and 4-22, respectively. Please ensure that you use the names specified in the following screenshots as that makes the walk-through a bit simpler to follow along.

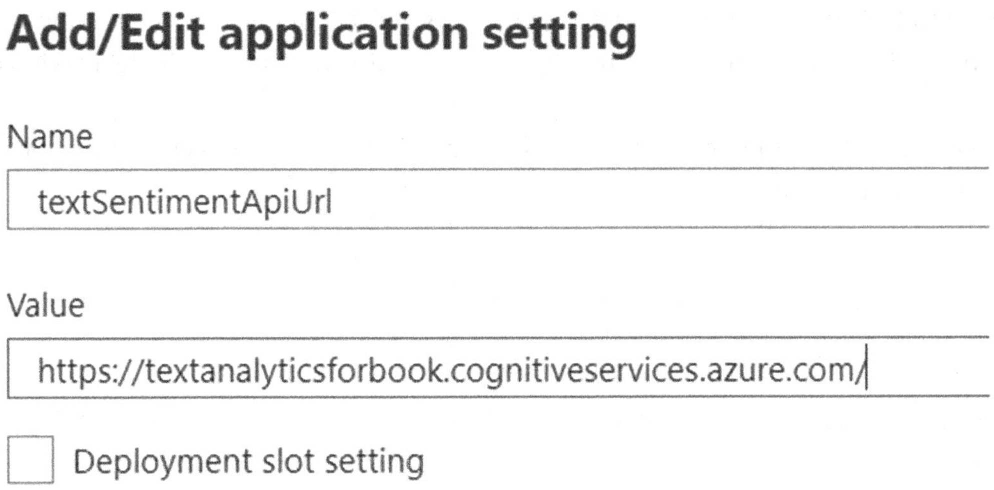

Add/Edit application setting

Name

textSentimentApiUrl

Value

https://textanalyticsforbook.cognitiveservices.azure.com/

☐ Deployment slot setting

Figure 4-21. Specifying the URL for the Text Analytics API the function will use

These two screens may not look that exciting, but they are absolutely critical to ensuring that your function connects to the right place using the right credentials.

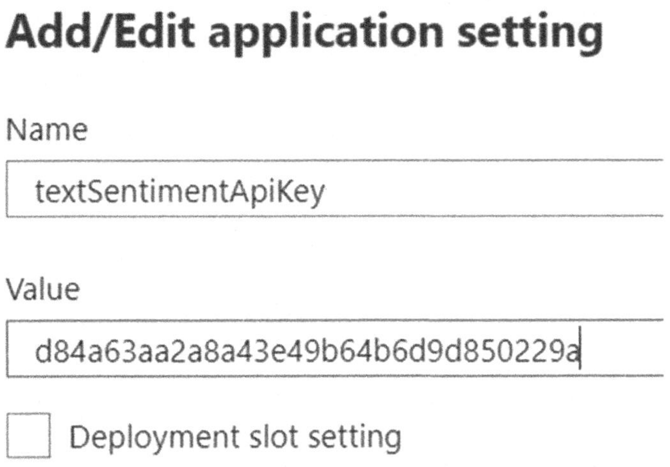

Add/Edit application setting

Name

textSentimentApiKey

Value

d84a63aa2a8a43e49b64b6d9d850229a

☐ Deployment slot setting

Figure 4-22. Specifying the API key used to connect to the Text Analytics API

Before we return to the logic app I referenced earlier in this chapter, let us add some code to our Azure Function so it functions as a function! If you don't want to dive any more deeply into the code, simply go to your Function App, click Functions, click LogicAppTrigger, click Code + Test, and paste the contents of LogicAppTriggerBody.csx (found in our GitHub repo) into run.csx. Put simply, it is basic C# code that categorizes a tweet as red, yellow, or green status based on the tweet's sentiment score. Feel free to change and tweak these values and designations as desired, but for now let us move on. We will do that by returning to the logic app you were directed to create in the previous section.

Back in the logic app, add a new step and Azure Functions as our action. Select Azure Functions as depicted in Figure 4-23.

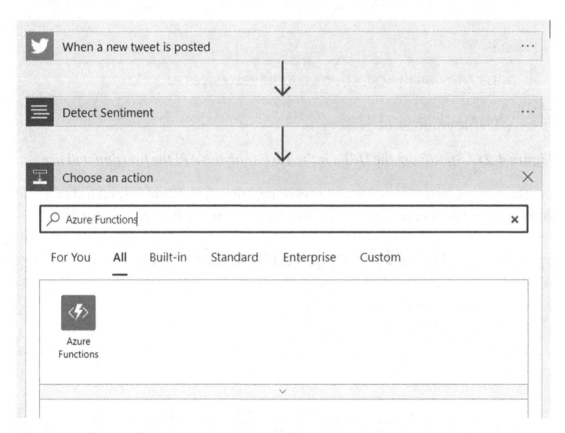

Figure 4-23. *Adding an Azure Function as a step in the Logic App*

We will then search for our Function App name (e.g., customercontact). After clicking that, we should be able to select our LogicAppTrigger function to include in our logic app's workflow as indicated in Figure 4-24.

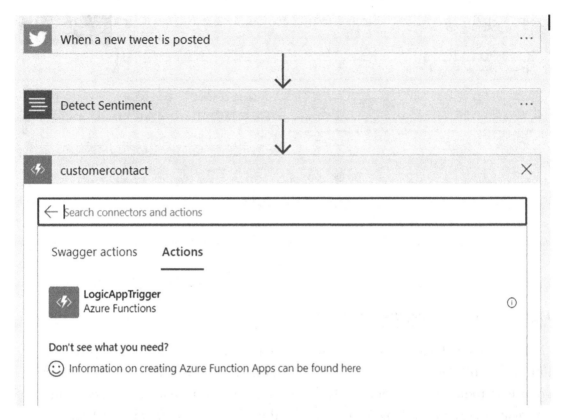

Figure 4-24. *Selecting the function to be executed in this step in the logic app*

That will bring us to the Dynamic content selector we have grown used to while working on these logic apps. In this particular case, we will select the sentiment score from our sentiment detection action earlier in the logic app workflow. This version of the Dynamic content selector is shown in Figure 4-25.

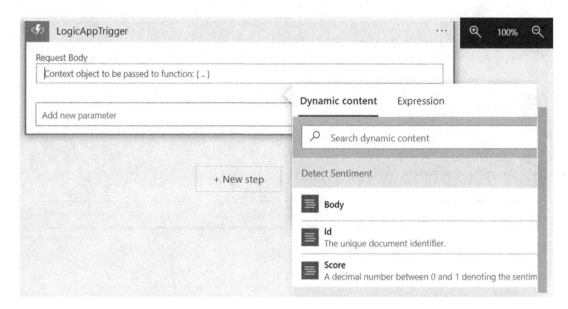

Figure 4-25. *Selecting the text to send to the Azure Function*

Having looped our function into our logic app, we want to add a new step, but this time a Control action. When we select Control, we will then select Condition as we are going to introduce some conditional logic into our logic app. That logic will result in an email to a resource asking them to contact the unhappy customer who is negatively tweeting about our company or its products. This Control/Condition selected is shown in Figure 4-26.

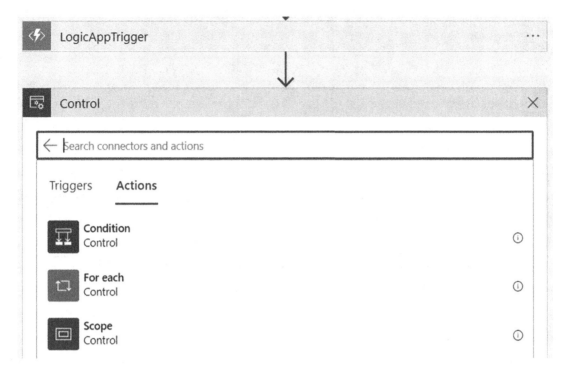

Figure 4-26. *Workflow for introducing a condition into our Logic App*

We will specify that our decision point here is when the body returned from the function (basically an HTTP Get) is RED. That means the sentiment score is quite low and requires communication from somebody to hopefully resolve the customer's or user's issue. Figure 4-27 shows how to specify the RED condition we wish to set up.

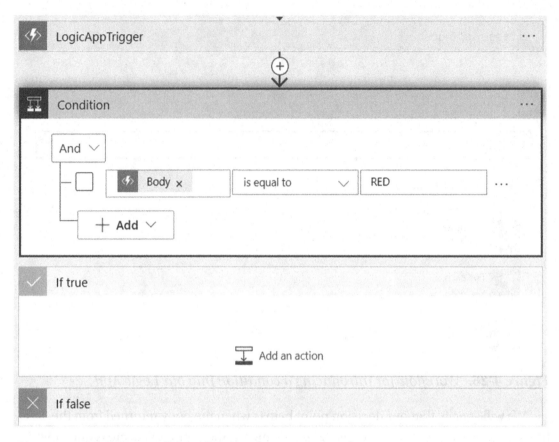

Figure 4-27. *RED condition shown after it is created*

After that, we need to choose how to send this email to our internal resource. While there are many options available to automatically send email within Azure, as I was building content for this book, I chose SendGrid as it seemed to be the easiest-to-use, most readily available email provider within Azure as of publishing time. I pre-provisioned a SendGrid resource. That said, if you prefer to use Microsoft 365, Gmail, or any other email provider to which you have access, substitute your provider of choice here. The screenshots won't be exactly the same, but understanding the workflow here is what it important.

Since we want to send this "alert" email if the tweet has been flagged as RED, let us select Add an action in the "If true" section of the condition, type SendGrid, and then click it. Figure 4-28 shows how to select the SendGrid action.

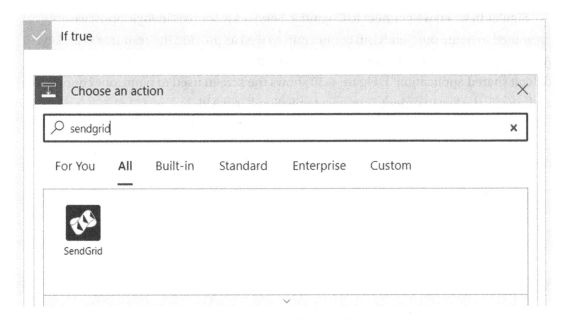

Figure 4-28. *Selecting the SendGrid action in the "If true" part of the condition*

Following that selection, find the Send email action in the list of SendGrid actions and click that. The Send email action selection is shown in Figure 4-29.

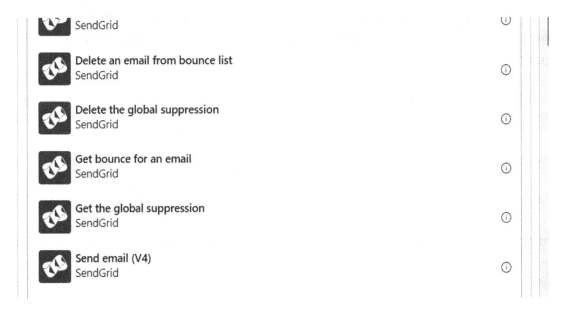

Figure 4-29. *Selecting the Send email action within the SendGrid action*

Similar to when we connect to Cognitive Services resources in logic apps, we are prompted to name our SendGrid connection as well as provide the resource's API key. After clicking Create, we will select the default authentication type for SendGrid ("Use default shared application"). Figure 4-30 shows the screen used to name our connection and specify the key in order to connect to the SendGrid API.

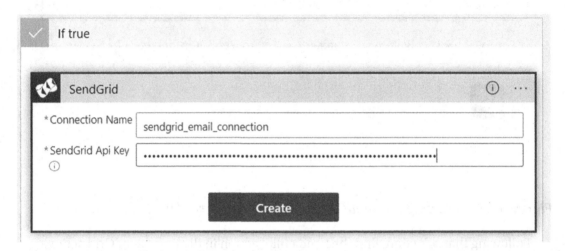

Figure 4-30. *Naming the SendGrid API connection and specifying the API key*

That brings us to adding parameters that you would expect of an email (From, To, Subject, Email body, etc.). Add those as appropriate and use my sample in Figure 4-31 as guidance.

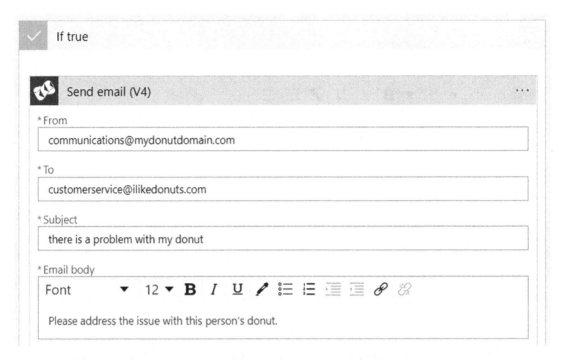

Figure 4-31. *Setting up the parameters for our alert email message*

Finally, if you wish, add the tweet text, Twitter username, or other tweet-related information as attachments so the customer service representative knows whom to contact and what issues they mentioned in their tweet. Figure 4-32 shows an example of how you would attach the tweet text information to the email message.

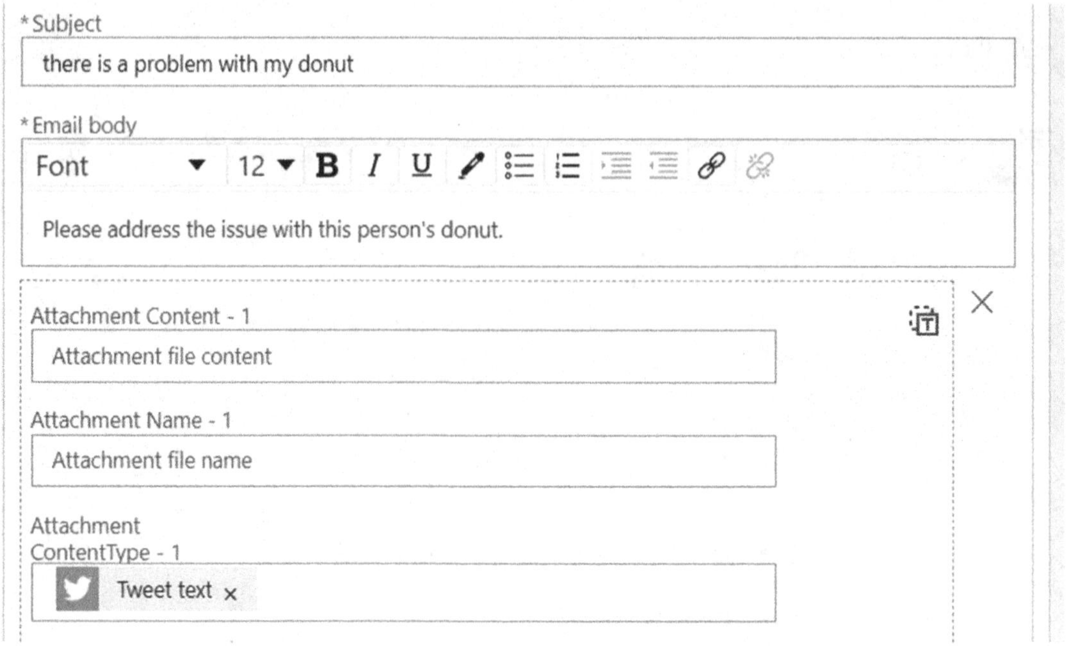

Figure 4-32. *Adding the text of the offending tweet as an email attachment*

Diversifying Communication with the Translator Text API

Thus far, we have tried to ease you into connecting to and communicating with Cognitive Services resources while utilizing a no-code option (like Azure Logic Apps) as well as an arguably low-code option (like Azure Functions). For those of you brand-new to Cognitive Services and/or new to software development in general, this probably makes sense. However, we know some readers really want to get into the meat of using application code to call a Cognitive Services resource and do something other than decide whether or not a tweet is positive or negative.

While we firmly believe that the last two sections provide a really solid framework to learn about these APIs (as well as the coolness of Azure Logic Apps and Azure Functions), this section of this chapter is for all of our readers who have worked through this section asking one or both of the following questions: "Why are there so many pictures?" and "Where is the code?" In this section, I will be referring entirely to the Program.cs file found in our GitHub repo (under Chapter 4). Minimal screenshots will be involved!

Using the Translator Text API to Translate Text

This console app is a basic demonstration of the capabilities of the Translator Text API found in the Cognitive Services Language API. As a brief summary, it takes input from the console, determines the language of the input, and then displays the output of its translation. Simple, yes, but the hope is that seeing the inner workings of this in this basic form will spark ideas on how you can leverage this ability within your apps or maybe entirely brand-new projects!

Main itself is the guts of the app but, in and of itself, fairly simple. Let us walk through the contents first, and then, if you would prefer an image in-line as opposed to looking back and forth between a screen and this book, there will be a couple in this section. Beyond all the Console.* lines that prompt the user the text they would like translated and then prompt the user to press a key to execute the translation, let's take a look at the two key lines within Main. First is the string defined as route in line 18. You'll note that that is basically the end of a URL – and it is. More on that in a bit as the end of this particular line is actually the most interesting part. You will see a series of things like "&to=en", and you may think that that is how you specify the languages you want the text translated into – and you would be right. If you are looking at the original version of the code in the GitHub repo, you will see the language codes for English, Italian, and German and then one language code that does not look familiar at all. If you want to earn a bit of knowledge, you can wait to find out what language that is by executing the code. If you don't want to earn that bit of knowledge, the language code tlh is the source of our fun fact in the following!

```
string route = "/translate?api-version=3.0&to=en&to=it&to=de&to=tlh";
```

Fun Fact   Did you know Cognitive Services can translate into Klingon?

Second, let us take a look at the await TranslateTextRequest in line 23. That is the actual call to the Cognitive Services resource itself. The first two arguments (subscriptionKey and endpoint) are the endpoint and API key referenced elsewhere in this chapter. While you certainly could provide this in the method call in clear text, that would be wildly insecure. For example, my code provided for this book would give you my API key and endpoint and likely increase my Azure bill quite substantially with everybody billing their experiment against my subscription! Instead, those are stored in environment variables. If you are not familiar with those within the context of .NET development, refer to the following URL for the latest guidance on how to set an environment variable in your operating system of choice:

```
https://docs.microsoft.com/en-us/azure/cognitive-services/cognitive-
services-apis-create-account?tabs=multiservice%2Cwindows#configure-an-
environment-variable-for-authentication
```

The third and fourth arguments are route (which we have already discussed) and textToTranslate (the text gathered from user input). Using await means this is an asynchronous call to the Translator Text API.

```
await TranslateTextRequest(subscriptionKey, endpoint, route, textToTranslate);
```

Now that we have discussed calling the TranslateTextRequest method, let us examine the highlights of that method rather than a line-by-line dissection in gory detail. The first section of the method (lines 55–61) is beginning to build the structure for an HTTP Post call sending a JSON payload. As discussed in many other places elsewhere in this book, communicating with the Cognitive Services APIs happens nearly exclusively via JSON payloads. This section ends with adding the API key (subscription key here) as a header to the request.

In my mind, I then break lines 63–86 into two sections. Lines 63–69 receive the string output back from the Cognitive Services API and deserialize the JSON. In an attempt to show more directly to the user, I have a Console.WriteLine(result) call in line 74, so the serialized JSON is displayed to the user. It is also the reason I have added four languages to the route variable – I believe it's much easier to understand the deserialization if you see multiple output sets returned from the API. If you would rather not see that now that you have seen the explanation of why it is there, feel free to comment out that line for prettier output. It is a console app, though, so "pretty" is a relative term here!

Lines 70–86 then loop through the deserialized results and output them to the console. One thing to note here is that the Console.WriteLine in line 77 shows off the language detection abilities of the Translator Text API (by outputting the detected input language returned from the API) as well as the confidence score returned by the API. That confidence score is the API's confidence (1 being most confident) that it has detected the correct input language.

Our hope is that this book finds a wide audience of folks interested in implementing AI using a wide variety of methods. While the code found in this section is certainly basic, we believe it provides both a nice on-ramp to calling Cognitive Services APIs via a .NET language (C# in this case) for those unfamiliar with application development and a solid understanding of the structure of the API call and the data returned from the API. This code (derived from a version of one of the Microsoft quick starts a couple years ago) has certainly sparked ideas and new development for us and now hopefully for you as well.

```
request.Method = HttpMethod.Post;
// Construct the URI and add headers.
request.RequestUri = new Uri(endpoint + route);
request.Content = new StringContent(requestBody, Encoding.UTF8, "application/json");
request.Headers.Add("Ocp-Apim-Subscription-Key", subscriptionKey);
```

Leveraging LUIS to Parse Text for Entities and Intents

This chapter largely emphasizes how we can "talk to" the Language API and, especially in the last section, what it looks like when it "talks back" to us. We have not really looked at the conversational abilities contained within Cognitive Services, though, so let us wrap up the technical portion of this chapter with a look at the Language Understanding Intelligent Service (LUIS). Later chapters will expand on the use of LUIS in chatbots and other conversational apps, but this section of this chapter should cover the basics to sufficiently prepare for the deeper dive into the capabilities of LUIS in later portions of the book.

Creating a Basic LUIS App

First things first, you connect to the LUIS portal by directing your web browser to www.luis.ai/. That front page provides you links to documentation, support, and other resources for learning about LUIS. There are also some basic demos and usage scenarios on the main page that provide a bit of insight into how you talk to LUIS, what it says back to you, and some usage scenarios for those capabilities. In this section, we will create two basic apps – one using a prebuilt domain/model to handle some basic restaurant reservation actions and one using a custom domain/model in which we attempt to order donuts.

133

When you sign in, however (via the link in the upper-right corner), you are taken to a different-looking part of the portal where you can create LUIS apps. That is where we will direct our attention for the remainder of this section.

As Microsoft pours investment and resources into AI development, we can see that manifested in the changes within Cognitive Services – it seems that every month all of the APIs have new abilities or improved abilities to perform tasks they already could do, but can now do at a more precise skill level. This rapid development, however, can often manifest itself in the UI of LUIS as well. Just in the period of time in which this chapter was being developed, the series of prompts to welcome your account to LUIS and prompt you to create a LUIS authoring resource changed three distinct times. To that end, we will forgo the screenshots that many of you are likely finding helpful and simply describe the flow of the process itself when you are creating an app.

Generally, the process will follow a route similar to the next few steps described. First, once you sign in, if you do not have a LUIS authoring resource provisioned, you will likely be prompted to create one. That screen may include an offer to be contacted with promotional offers as well as a checkbox to agree to the Terms & Conditions of provisioning a LUIS authoring resource. You may be offered the choice of using a trial authoring resource, but it is recommended that you create your own authoring resource. When you click the button to create your new LUIS authoring resource, you should be prompted for the name of your new authoring resource, the subscription where it will reside, and the resource group inside that subscription where you would like to place the authoring resource.

That will likely direct you to a splash screen that may look similar to Figure 4-33. From that screen, you should click the "Create a LUIS app now" button to dive into the app creation process itself.

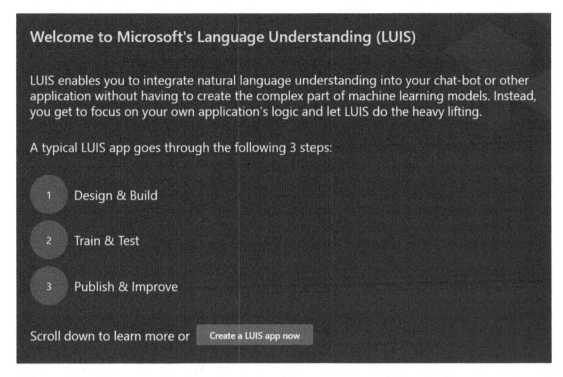

Figure 4-33. *Introductory screen in LUIS for a new conversation app*

While you can import a JSON document or LU (Language Understanding) model, that is not where we would start to gather the basic understanding of this that we seek to start off here. To that end, select the choice for "New app for conversation" so we can use the GUI to build our new conversation app that can be used by other resources like bots. Figure 4-34 shows us how to make that selection.

Figure 4-34. *Selecting "New app for conversation" to begin LUIS app creation*

Unsurprisingly, when we click to create an app, we will arrive at a screen whose look we are beginning to grow accustomed to. We will select a name for our new chat app, a culture, and a description. It is important to note the text under the following Culture field – the culture selected there is the language in which the app converses, which may be different than the language present in the interface itself. This may be because your customers and LUIS developers speak different languages or of any number of any other reasons, but it is useful to have this option, and we wanted to make sure to call it out. The final field is a prediction resource (used in training and testing your model), but it is for more advanced LUIS users, so we will not create or select one of those for now. Figure 4-35 shows us an example of the "Create new app" screen.

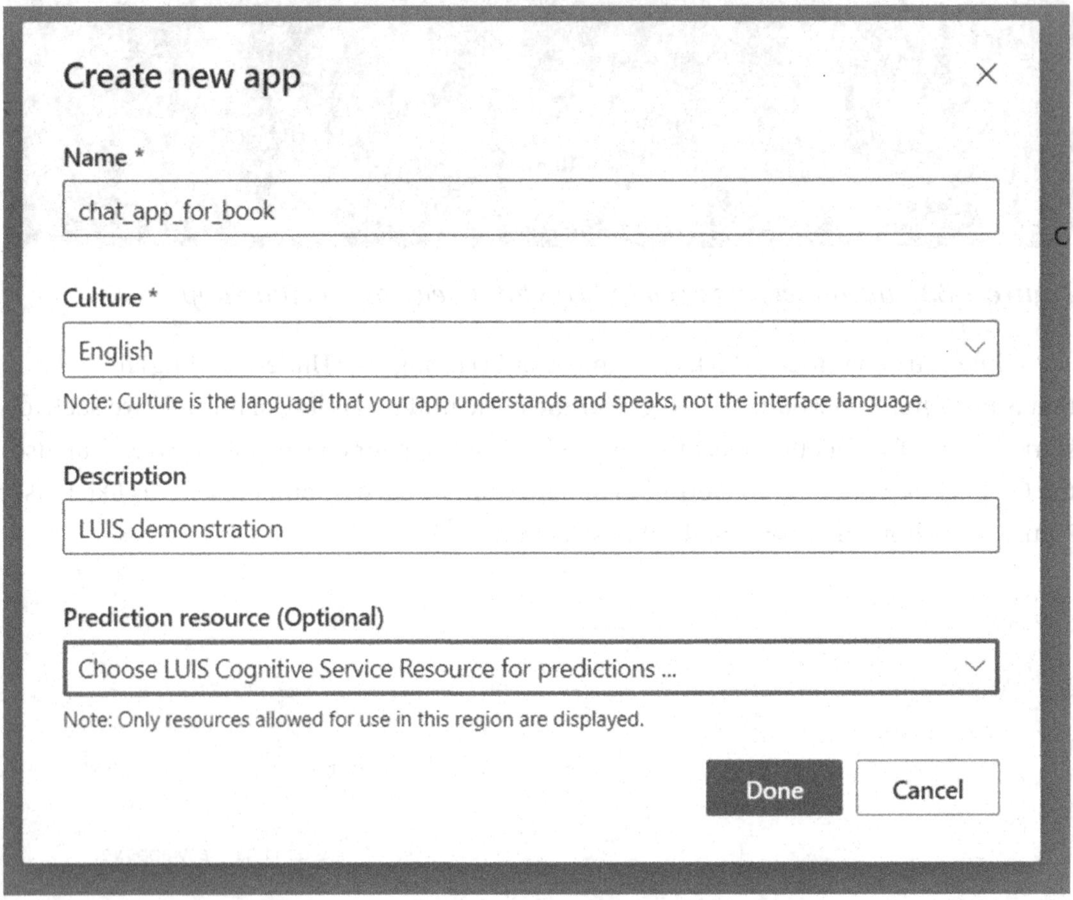

Figure 4-35. *Specifying basic details for our LUIS app*

When we click Done here, we should be taken to a list of prebuilt models from Microsoft that will allow us to take a quick look at some basic things you can do with LUIS straight out of the box (so to speak). For our purposes here, we are going to click the Add domain button under RestaurantReservation. If we were not taken to a list of prebuilt domains, all prebuilt domains can be accessed in the Prebuilt Domains area under App Assets at the far left of the screen (see in the following). Figure 4-36 shows how to navigate to Prebuilt Domains.

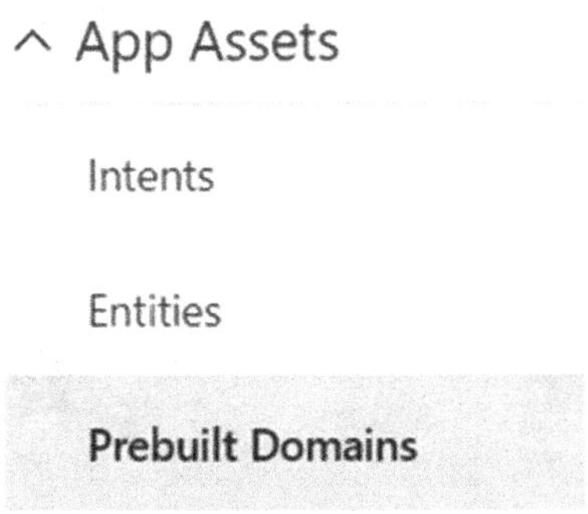

∧ App Assets

Intents

Entities

Prebuilt Domains

Figure 4-36. *Navigating to Prebuilt Domains within LUIS*

However we arrived at adding the prebuilt RestaurantReservation, our screen should now look something like Figure 4-37.

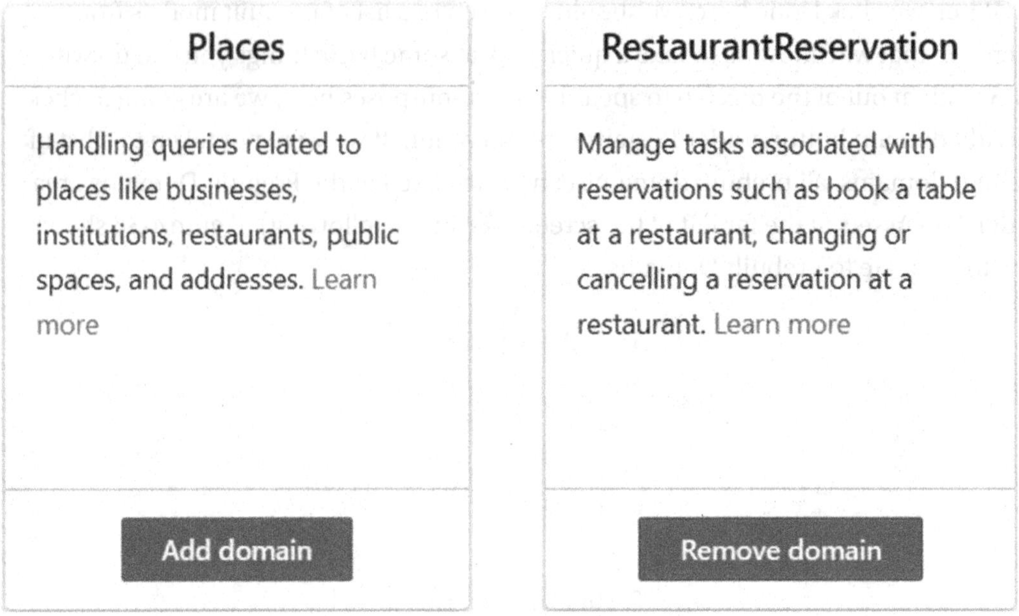

Places	RestaurantReservation
Handling queries related to places like businesses, institutions, restaurants, public spaces, and addresses. Learn more	Manage tasks associated with reservations such as book a table at a restaurant, changing or cancelling a reservation at a restaurant. Learn more
Add domain	Remove domain

Figure 4-37. How RestaurantReservation will look after you have added it to your app

Now that we have added this prebuilt domain, let's click Entities in the upper-left portion of the screen. Here you will see a variety of entities that may be a component of a restaurant reservation: its address, the type of meal(s) it serves, its name, the number of people in the reservation, etc. While entities are not required to be built, understanding what they are and their three different types is critical to your LUIS app facilitating a functional conversational experience. Before we get into the types, though, let's understand what entities are within the context of LUIS. Entities are data you want to extract from the utterance. The utterance is the thing the user of your app says to LUIS. If your user says to your app, "I want two donuts," the entities in that utterance are defined as the number (2) and the thing (donut). Referring back to our list of entities, you will note that you see two distinct types of entities: "list" and "machine learned." While there are other types, list and machine learned are most typically encountered and definitely the two most used types for users new to LUIS. List is a defined list and not learned through training the machine learning model. For example, RestaurantReservation. MealType is a hard-coded list of the types of meals this fictitious restaurant serves. RestaurantReservation.NumberPeople is noted as a machine learned type, and that is because, as this model was built and trained, the number of people in the utterance was identified as this particular entity and the model continues to build on that definition as it is trained.

After understanding the prebuilt entities in this model, let us quickly browse the list of prebuilt intents as well. Specifically, let us review RestaurantReservation.Reserve. That is the intent that is identified when an utterance contains something along the lines of "I would like to reserve table of 4 at 6 PM." Clicking the entity not only shows us examples of utterances identified as targeting this intent, but it also shows graphical examples of the entities identified in utterances. These are the underlined labels seen in the following screenshot and worthy of further exploration in the portal as your time allows. Figure 4-38 shows some of the associated details and examples for RestaurantReservation.Reserve.

RestaurantReservation.Reserve ✎

Machine learning features   ?

+ Add feature

Examples   ?

✓ Confirm all entities   ⬚ Move to ⌄   🗑 Delete   ···   🔎 ≡ View options ⌄

Example user input	Score
Type an example of what a user might say and hit Enter.	

could you help me reserve a 5 - star restaurant room
                       RestaurantRe...   ⋮

make a reservation at a restaurant nearby
                                   Restaura...   ⋮

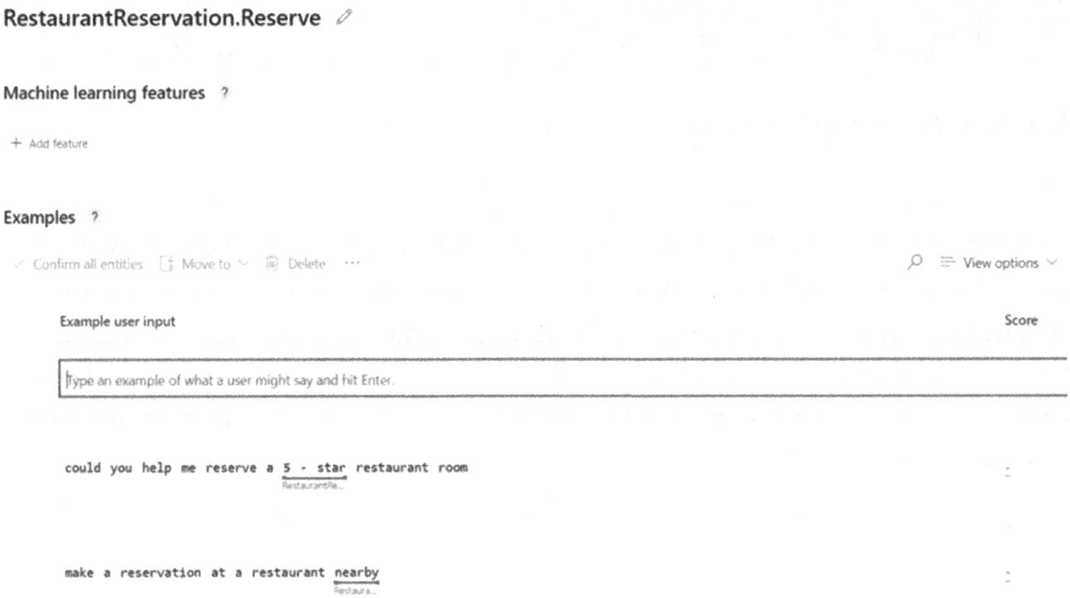

Figure 4-38. *A look at the details of RestaurantReservation.Reserve*

In the beginning of this section, we said that we were going to talk to something in this section, and so far we have done a lot of talking about something and not talking to something. It is time to change that. If you look to the upper right of your app screen, you should see a Train button with a red light on it. That basically means that you have not yet trained your model so you can talk to your app. Click that button (as shown in Figure 4-39), and the training process will begin.

Figure 4-39. Training our model

Once the training process has completed, you will see that red light turn green, and that means your app is ready to test. Click the Test button and start the conversation! Figure 4-40 shows us what it should look like when the training process is complete.

Figure 4-40. Model training complete – they grow up so fast!

At this point, you can type anything you want and investigate the response from LUIS. To keep our examples cohesive and concise, let us make sure that it is correctly identifying utterances that want to use the RestaurantReservation.Reserve intent. To that end, we have typed "I want to reserve this restaurant" into the Type a test utterance... field and received the following response. It has correctly identified the intent, and it is fairly confident that it has identified the correct intent. Figure 4-41 shows us an example of testing our LUIS model.

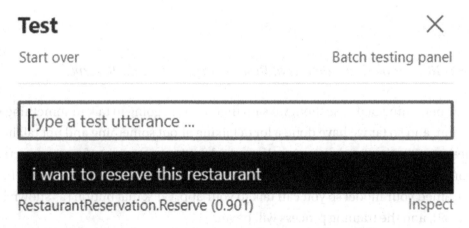

Figure 4-41. Beginning the conversation with our LUIS model

That is a good start, but there is only basic information there. If we click the Inspect hyperlink, we can see every bit of feedback we received from LUIS. It lists the user input (which of course we already knew), the top-scoring intent (which is the one returned to the screen under our utterance), and other more advanced information that would be of interest to us if we were trying to create new entities or entities in different (and more advanced) types than the ones we have already discussed. We want you to know that this information is here when you click Inspect, but, for now, our journey into the basics continues as we click the X in the upper-right corner and move beyond this screen. Figure 4-42 shows an example of the X we should be clicking to exit.

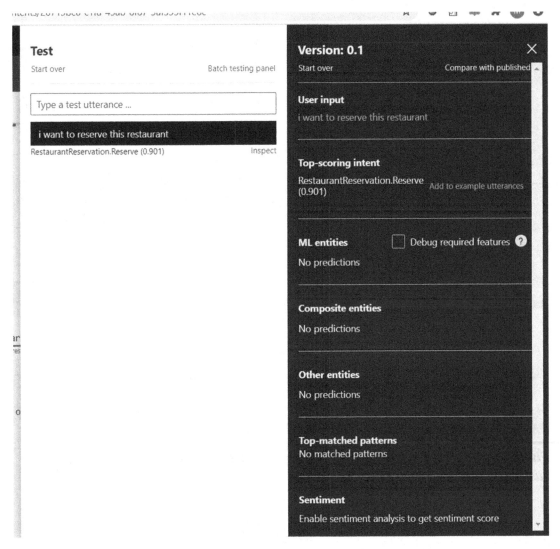

Figure 4-42. *Click the X to exit*

While prebuilt models, and the intents and entities they contain, can be great tools to begin playing with LUIS and understanding its components, the real fun comes when you begin creating custom apps with custom models. When teaching on this topic, that is often when a light bulb really comes on for the people in attendance. For our custom app, we will follow the same steps we did to create our "chat app," but let us give it a descriptive name so it at least indicates to us that it is the app we have built to play with custom things that we are building and training. Figure 4-43 shows us how to get started creating our custom app to use our custom model.

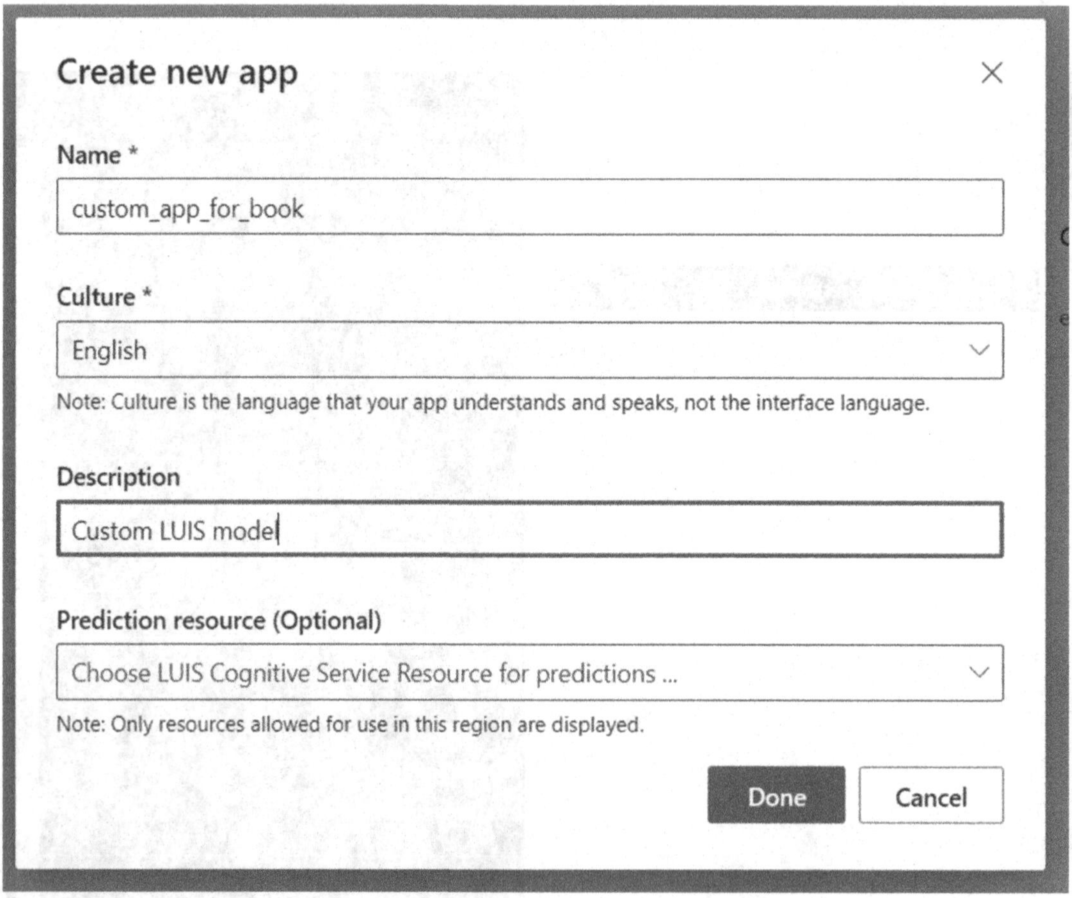

Figure 4-43.   *The "Create new app" screen in LUIS that will allow us to use our custom model*

We have an app but no intents. An app with no intents will make a very dull conversationalist indeed! To create an intent, we will click Entities in the upper-left corner again, but, instead of reviewing what is already there, we will click Create because there is in fact nothing there yet.

Our first example will be OrderDonut because, typically, the first thing anybody wants to do with a donut shop (even one that does not exist) is order a donut. That brings us to a screen similar to what is shown in Figure 4-44.

Intents ?

+ Create   + Add prebuilt domain intent   ✑ Rename   🗑 Delete

○	Name ↑	Examples	Features
	None	0	+ Add feature

Figure 4-44. *Creating our first intent*

Next, in Figure 4-45, we will name our intent.

OrderDonut ✎

Machine learning features ?

+ Add feature

Examples ?

✓ Confirm all entities   Move to ∨   🗑 Delete   ···

Example user input

Type an example of what a user might say and hit Enter.

There are no utterances found.

Figure 4-45. *Naming our first intent*

Our intent has a name but nothing more than that. To get things started, we need to provide LUIS some basic utterances that we think would trigger this intent in the conversation. We have provided few examples that are basically long-winded ways of saying "I want a donut." We can recommend providing that as an example as well (but it may make you actually want a donut). Figure 4-46 depicts one example of an utterance for this intent.

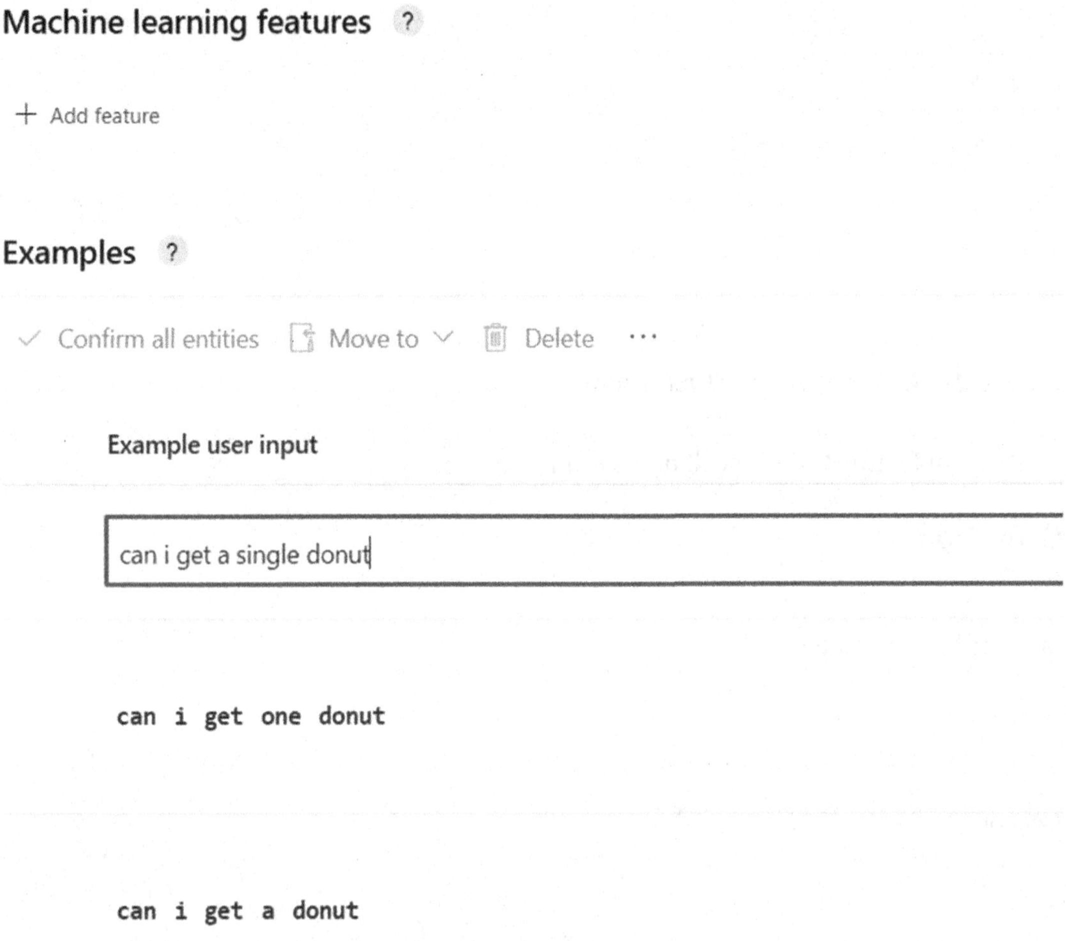

Machine learning features   ?

+ Add feature

Examples   ?

✓ Confirm all entities   Move to ∨   🗑 Delete   ...

Example user input

can i get a single donut|

can i get one donut

can i get a donut

Figure 4-46. *Providing some example utterances*

While the OrderDonut intent is likely what our customers ultimately want, we want to provide a more conversational experience to them than simply the ability to order a donut from our pretend donut shop. To that end, let us create another intent that will help LUIS understand that the customer is greeting it – similar to what will ultimately lead to the more conversational interactions detailed in Chapter 7 later in this book. You are welcome to provide it many examples of greetings – a couple of our suggestions are in Figure 4-47.

GreetingsAndSalutations ✎

Machine learning features ?

+ Add feature

Examples ?

✓ Confirm all entities   📷 Move to ∨   🗑 Delete   ⋯

Example user input

| Type an example of what a user might say and hit Enter. |

wazzup

what ' s up

Figure 4-47. *Teaching the model to greet customers in a very informal way*

You can also provide an intent that helps LUIS understand that the customer agrees with the question it has been asked. There are many examples of this, but creating this intent could look similar to the example in Figure 4-48.

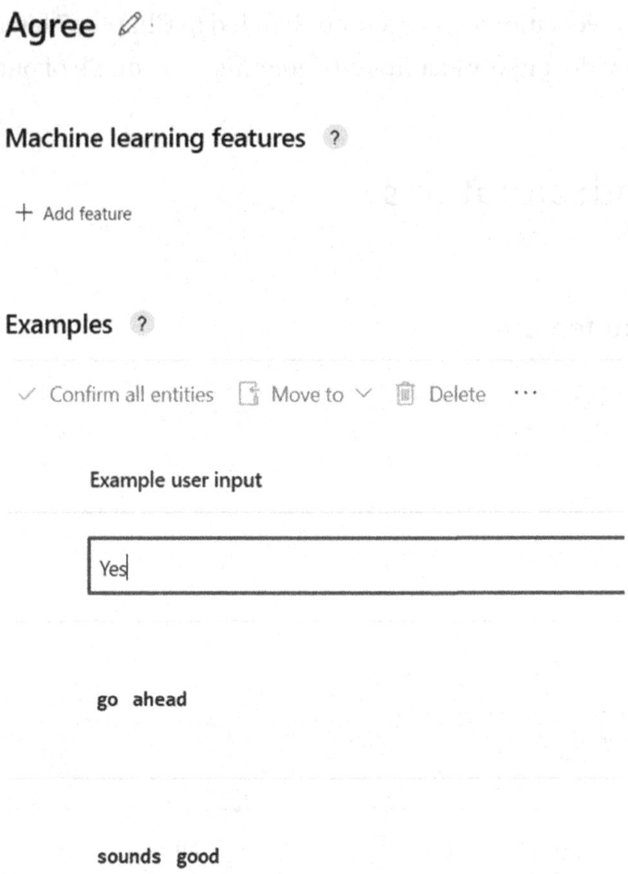

Agree ✎

Machine learning features ?

+ Add feature

Examples ?

✓ Confirm all entities   ⎗ Move to ∨   🗑 Delete   ⋯

Example user input

Yes|

go ahead

sounds good

Figure 4-48. *Teaching the model how to agree with a customer*

Finally, in pursuit of training the model, it is recommended that you create an intent called None (or a similar term) and put intents in there that are completely irrelevant to the app's task at hand. This can have a subtle effect on the training of the model, and it can also be quite fun to come up with silly examples of complete non sequiturs as well! Figure 4-49 shows a few of our attempts at these.

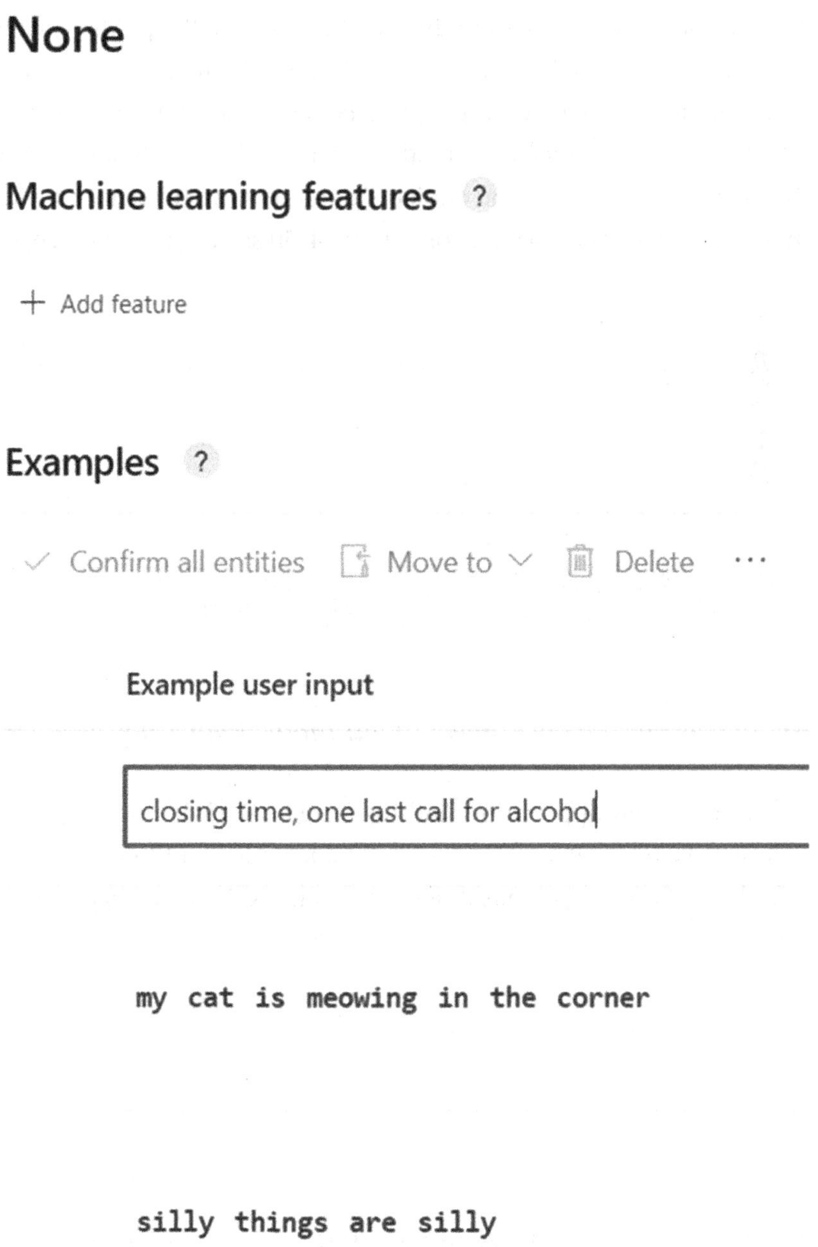

None

Machine learning features ?

+ Add feature

Examples ?

✓ Confirm all entities   Move to ∨   🗑 Delete   ⋯

Example user input

closing time, one last call for alcohol

my cat is meowing in the corner

silly things are silly

Figure 4-49. *Teaching our model random things*

To wrap up our basic LUIS app, train this one as we did in the previous example. Once that "light" is green, click Test, and we will find out if our app is taking those initial steps toward understanding what we are saying. For starters, let us tell it something random like "pizzas are awesome." It correctly identifies that as the None entity, although it is not very confident of that choice. Remember, we have not provided the model much data to use to make a decision. Figure 4-50 shows us our success.

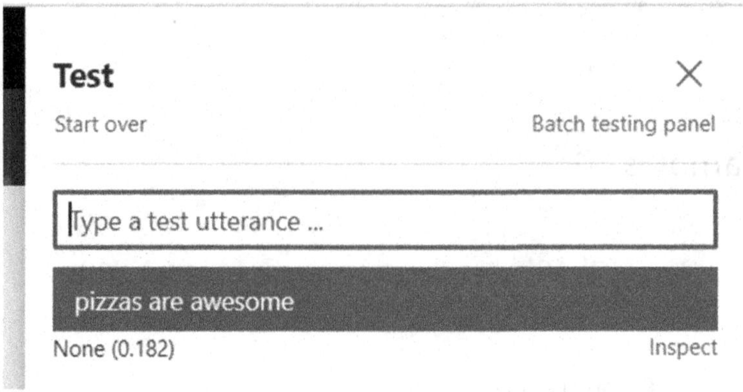

Figure 4-50.   *Pizzas may be awesome, but our model knows it is about donuts.*

Next, let us test our new app with something a customer is likely going to say to it: "I would like a donut." We do that, and we see that it correctly identifies the OrderDonut entity and is fairly confident that it has that correct. Figure 4-51 shows you an example of how to test this.

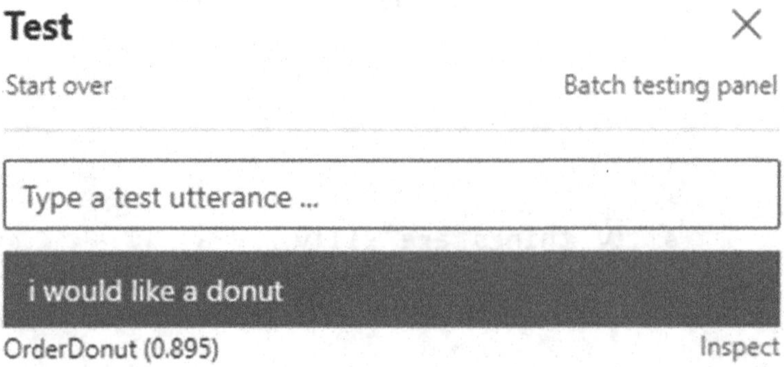

Figure 4-51.   *Speaking to our donut model about donuts*

To wrap up our testing (and this section), let us see if we can throw it off by ordering something that is not a donut. Remember, the intents we specified in OrderDonut all mentioned donuts. Is the model smart enough to correctly classify our utterance is non-OrderDonut or not? Let us tell it "I would like a pizza" and see what happens.

Surprisingly or not, depending on your perspective, our final test shows that our trained custom model was not to be fooled and did in fact correctly classify our request to order a pizza from the invisible donut shop as the None entity rather than OrderDonut. Figure 4-52 shows that the model was not fooled.

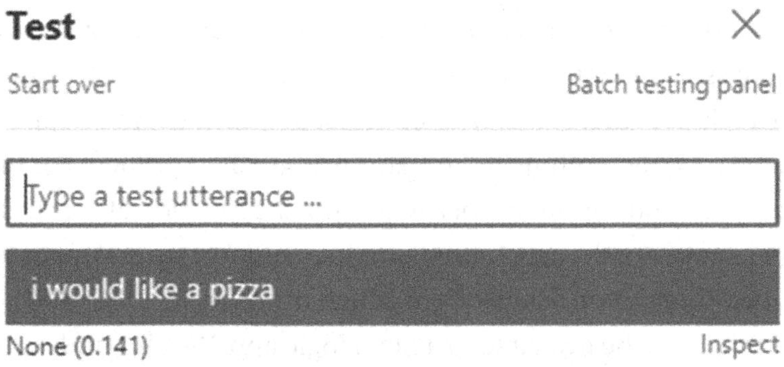

Figure 4-52. *Our trained model acting just as we taught it*

Summary

Much of this chapter may have seemed introductory – that was intentional. Parts of Chapter 6 and all of Chapter 7 will build on the concepts discussed in this chapter. Without understanding how Language resources are provisioned, how to communicate with them, and how they communicate back, your work and learning in the later chapters would be built on a less than sturdy foundation. Let us briefly summarize each section before you move on to learning about the Speech API in the next chapter.

Sentiment Analysis

Sentiment analysis is but one feature of the Language API, but it is often the entry point (or perhaps gateway drug) for deeper exploration of everything the API can do. For at least one of the authors, playing with sentiment analysis has led to several AI-related projects. Just because anything (or everything) in this chapter, or indeed in the entire book, is not germane to your current career does not mean it will not lead you to your future career. Using an Azure Logic App introduces you to a no-code way to engage with the Language API (with a side dish of learning how useful logic apps can be). After reading that section, it is our hope that any technical professional will be able to build logic app–based solutions that utilize sentiment analysis or other functions of the Language API – even if that technical professional does not want to write any code at all.

Azure Function Integration

We built on our knowledge of logic apps from section "Create an Azure Logic App" and incorporated an Azure Function in our workflow. In this case, our Azure Function was simply used to classify the status of a tweet as red, green, or yellow. The function was called from the logic app, that status was returned, and then the logic app sent an email if the status was red (and did nothing if the status was green or yellow). For those who did not appreciate the no-code approach of a logic app, the function allows us to introduce code (C# in our example, although Azure Functions support several other languages) into our Cognitive Services workflows. That allows us to create richer and more robust logic than may otherwise be available in a plain logic app. We also understand how to create an Azure Function in case we would like to forgo a logic app entirely when calling Cognitive Services APIs.

Translator Text API JSON Deserialization

After introducing a bit of code in section "Driving Customer/User Interaction Using Azure Functions and the Language API," section "Diversifying Communication with the Translator Text API" dove headlong into the gory details of what a Language API call looks like without the friendly wrapper of a logic app and/or function. As discussed throughout the book, communication with Cognitive Services APIs rides the rails of a JSON payload, and this section gave a detailed look as to how to build that JSON payload in a C# console application and then display the deserialized output from the Translator Text API. We review the Translator Text API's language detection capabilities as well as its ability to translate from one language to many others.

LUIS Training and Components

Finally, after talking about a lot of things and learning how to talk to the Translator Text API, we ended this chapter on the Language API learning how to create a conversational experience with LUIS. While Chapter 7 will take a deeper dive into LUIS and help us to understand how to wrap our prebuilt or custom LUIS models in a chatbot experience, in order to get there, we first needed to learn the basic components of a conversation in LUIS, how to create an app and its components, and how to test it. We are also likely a little frustrated that we haven't been able to order a donut even though we built an app to do that! While we end this chapter on a conversational note, that is a solid segue into Chapter 5 where we will learn more about language that we can hear: speech via the Speech API.

CHAPTER 5

Speech Services

The Speech service has become a trendy topic with its flexibility for application integration. Nowadays, people not only send text messages to each other – they also feel comfortable sending audio messages to each other. Audio messages mean it's possible for someone to send a message when their two hands are otherwise occupied, like when someone is commuting to work. From there, think about more scenarios where hands-free communication is needed, like trying to play (or stop) the cooking tutorial video while learning how to cook, picking up a phone call while you are busy with household duties, or on the way to the office with your coffee in one hand and your friend's coffee in the other.

Using the Speech service allows you to create a better user experience – with audio, you can more easily convey emotion than you can with a text message. The Speech service also improves accessibility for differently abled groups of people, particularly people with vision challenges. As many of the other Cognitive Services do, the Speech service seeks to democratize AI and AI-powered technology. It sounds interesting and practical, doesn't it?

In this chapter, we will start with an overview of the Speech API, move to covering Text-to-Speech and Speech-to-Text services in greater depth, and then ramp up our knowledge by diving into the Speaker Recognition service and more advanced abilities of the Speech service. For reference, the demos and code highlighted throughout this chapter were developed in an environment using Python 3.8, macOS, and Visual Studio Code. Let's get started!

© Alicia Moniz, Matt Gordon, Ida Bergum, Mia Chang, Ginger Grant 2021
A. Moniz et al., *Beginning Azure Cognitive Services*, https://doi.org/10.1007/978-1-4842-7176-6_5

Speech API Overview

The Azure Speech service is part of the suite of Azure Cognitive Services. For developers who are new to working with audio and/or machine learning, you still can use the RESTful call to interact with the Speech service. With these RESTful calls, you can enable this speech functionality within your application.

For example, have you wanted to give audio or video content accurate captions so it's accessible to a wider audience? Have you wanted to provide a better user experience for users of your application by offering them voice interaction? If you are working on behalf of a business, have you wanted to gain insights from customer interactions to improve your applications and services?

If any of these ideas sound compelling, then the Speech service will be a must-have to help you enrich your application and provide a better user experience. The Speech service has been used in travel, education, logistics, manufacturing, and many other industries. Throughout this chapter, you will learn more about the variety of use cases of the Speech service along with how it works together with the APIs within Language to do things like translate the content to another language. First up, let us get into the use cases for the Speech service.

Applied Cases in Real Industries

When people talk about AI and machine learning, they talk about how fancy face recognition and object detection work. People talk about how much time and money they save with the new technology. Figure 5-1 shows the three common application use cases of the Speech AI service: navigation, education, and accessibility.

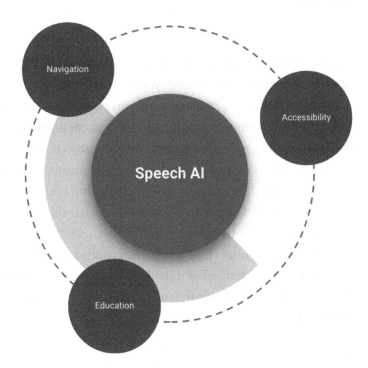

Figure 5-1. *Three common application cases of Speech AI: navigation, education, and accessibility*

Navigation

With a camera and an augmented reality app, you can guide visually challenged people to visit a new place with the Speech API helping them by transcribing the 3D world via the application. The availability of affordable 3D depth sensors brings new opportunities for these people and others to explore the world with the combined powers of 3D sensors and the Text-to-Speech service.

As the Vision APIs are also discussed in this book, think about the possibility of having an "eye-free" yoga class where you follow the audio instruction and your given audio cues adjust your posture based on what a posture alignment service is seeing through a commonly available camera. You would be able to enjoy the yoga class without sneaking peeks to your screen to see every single subtle movement the teacher makes.

It's also possible to increase content accessibility for users with different abilities, provide audio options to avoid distracted driving, or automate customer service interactions to increase efficiencies.

Education

We've likely all used an app or a website that has a speech component. Not only does this make for a richer student experience but it makes your content both more interesting and more accessible to a broader audience. By using a code-free tool such as Speech Studio, you can provide that richer audio content without any development knowledge necessary.

While the bulk of this chapter will discuss code-focused ways to use the Speech API, it is important to understand that there are other methods to interact with the API that ask far less technical knowledge of the user than writing Python code.

By utilizing the Speech API and Speech-to-Text API, you are also able to provide speech therapy services more easily to those in need of them. One of the authors of this book sat for speech therapy as a young child using a microphone and a cassette player. While that can definitely not be described as a rich experience, modern tools like these APIs can provide an interactive experience that will help bring a variety of students to speech and linguistic proficiency.

Accessibility

Finally, the Speech service, in cooperation with other parts of the Cognitive Services suite, is able to powerfully make accessibility a reality for so many who face an accessibility struggle at work, at home, and out in the world. Using the Speech-to-Text and Text-to-Speech abilities of the service combined with the translation powers of both Speech and Language, you can bring the world to your users, students, customers, etc. Whether it's the world near to them or the world far from them, they can experience it in vibrant ways regardless of their language, location, or abilities.

Along these lines, there are two programs from Microsoft that utilize and facilitate the power of AI to help users and developers with their accessibility journey. Disability Answer Desk is a Microsoft-provided service desk. Customers with disabilities can get support for Microsoft Office, Windows, and Xbox Accessibility as well. This service desk includes product issues, accessibility questions, and use of assistive technologies. The AI for Accessibility program is a Microsoft grant program for AI accessibility, which brings the power of AI to amplify human capability for the more than one billion people around the world with a disability. It can provide the means and technology to bring people experiences that they may have been missing at this point in time. To learn about the AI for Accessibility program, please visit this website for more information: `www.microsoft.com/en-us/ai/ai-for-accessibility`.

Interacting with the Speech APIs

Now that we have reviewed some use cases for the APIs, this is a good time to review the five different ways you can communicate with those APIs.

First up, let's briefly discuss the Speech CLI (command line interface). Using the CLI is often a very good choice when working with advanced service requests or developing custom behavior (some of which is noted later in this chapter). Despite what you might think, the CLI can interact with the audio input from the microphone, prerecorded audio files, or text files directly.

Second, you can use the Speech SDK (Software Development Kit) to interact with these APIs. There are seven different programming languages to choose from that are supported by the Speech SDK: C#, C++, Go, Java, JavaScript, Objective-C, and Python. The Speech SDK exposes many features from the Speech service, but not all of them. If you need a specific feature that is accessible via the Speech SDK, we recommend to cross-check the REST APIs for more feature support.

Third, you can use the Speech Devices SDK when you have an application that needs to run on Windows, Linux, or Android devices. As with many Azure services, there are some features that are limited in certain Azure regions, so it is always best to check the online Cognitive Services documentation for what is currently supported where as that is constantly updated. This is a good thing to keep in mind when you are developing and deploying things for strictly one or two reasons. You will not want to get too far into the development effort before you realize that the regions you are using are not ready to support these features! The supported programming languages for this SDK are Java and C++. The devices should work with ROOBO Dev Kits as well as the Azure Kinect DK.

Fourth, Speech Studio is made available as a customization portal for the Speech service on Azure Cognitive Services. It supplies you with a set of code-free tools to interact with the Speech model. Within the studio, you can prepare your custom models, test them, and then monitor them post-deployment. Speech Studio also provides you with custom models to work with your available audio speech data along with the ability to dive deeper into preparing and testing your models.

Finally, there are the REST APIs to which you can make RESTful calls. Within the Speech service, there are APIs for Speech-to-Text, Text-to-Speech, and Batch Transcription and Customization. With the REST APIs documentation found on the Cognitive Services website, you will be able to interact with the API from a REST call without the SDKs mentioned earlier. It is important to point out that, despite so many things in the technical world being rather poorly documented, the Cognitive Services documentation is

157

frequently updated and full of examples as well. That documentation can be found here: https://docs.microsoft.com/en-us/azure/cognitive-services/speech-service/ rest-speech-to-text. If you are looking for the more advanced features like copying models, transcribing data from a container, getting the logs per endpoint, etc., you can find all of that advanced functionality available within these API calls.

For the rest of this chapter, I will focus on utilizing the Speech SDK using Python. The SDK provides a nice middle ground between code-free tools like Speech Studio and code-heavy methods like API calls. After practicing with the Speech service using the code samples, I will introduce some common voice datasets. I encourage you to use them for continued learning as well as learning how to dive deeper into your audio data to obtain more knowledge and insights. From there, you will be able to try many different scenarios and, using the use cases mentioned earlier as a starting point, think about all the interesting (and potentially helpful) ways you can use the Speech service.

Text-to-Speech and Speech-to-Text

Given how many times we've referenced them already, it will not surprise you to learn that the Text-to-Speech and Speech-to-Text APIs are the most frequently used within the Speech service. These two APIs provide the possibility of working hand-in-hand and uniting audio and text communication from end to end. While many of us are likely familiar with using text-to-speech and speech-to-text abilities in our phones and other devices, it is likely that you are reading this book because you want to learn how to use them on your own.

To begin that journey, we are going to create a Speech service within Azure and get familiar with the Speech service interface on Azure Portal. Let's take a look at how we can do that.

Create a Speech Service

In this tutorial, a new Speech API will be created via Azure Portal:

1. Go to Azure Portal: https://portal.azure.com/#home. Once it loads, type "cognitive services" in the search bar. Click the Create button to create a new resource. Search for "speech service" in the Marketplace search bar (as seen in Figure 5-2), select Speech (the one provided by Microsoft), and you should find the Speech service presented similarly to how you see it in Figure 5-2.

Figure 5-2. *The Speech service in Microsoft's Cognitive Services marketplace*

2.   In this step, we will set up the service name with free text, so type a
     name that will help you identify it within your Azure resources. Next,
     we need to select the subscription where we want to deploy this
     Speech service. Following that, select the location of the service. It
     can be the closest location to your geolocation, or you may choose
     the one that is closest to your user or customer. The next setting is
     the pricing tier. Using Speech-to-Text as an example, each of the
     services has two pricing tiers, free (F0) and standard (S0).

The free tier of online transcription does limit the number of concurrent requests
as you would expect. You can have one concurrent call for both a base and custom
model. When using the standard tier, you can have 100 concurrent requests for the
base model and 20 concurrent requests for the custom model. Within the standard
tier, you can also have an adjustable model for both base and custom model modes.

---

3.   Lastly, select (or add) the resource group for this API. A resource
     group stores metadata about the resources for an Azure solution.
     If you have an existing resource group for your other Cognitive
     Services projects, you may use that here. If not, it is perfectly
     fine to create a new one for this project that is new to you. The
     parameters we have discussed so far are shown in Figure 5-3.

Home > Speech services >

Create

Speech   ×

Name *

speech-service

Subscription * ⓘ

Microsoft Azure Sponsorship

Location *

(Europe) West Europe

Pricing tier (View full pricing details) *

Standard S0

Resource group * ⓘ

book-ai

Create new

Create   Automation options

Figure 5-3. *The parameter setting on the Create page of the Speech service in Microsoft's Cognitive Services marketplace*

4. Typically after 1–2 minutes, Azure Portal will show the status change from "Deployment is in progress" to "Your deployment is complete." The details of the deployed service will be shown in fashion similar to Figure 5-4.

5. On the left-hand side of the navigation bar, you will see Overview, Inputs, Outputs, and Template. By selecting Template, the page will show the deployment setting as a JSON file, which preserves the input parameters. You may use this JSON file for script deployment.

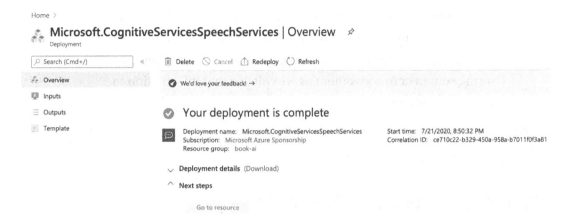

Figure 5-4. *The deployment status of the Speech service in Microsoft's Cognitive Services Speech service*

6. When the "Go to resource" button on the Overview page turns blue, click it to visit this deployed service. In the Overview tab, you will see Get Started, Discover, Develop, and Deploy. Select the Discover tab, and then you will be able to have the first interaction with the API you just created. The page should look like Figure 5-5.

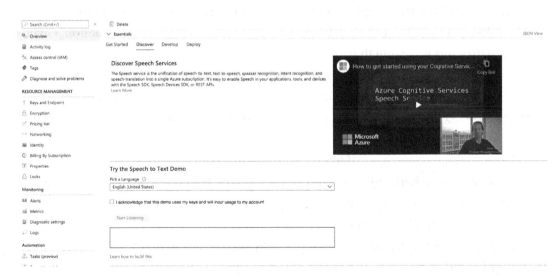

Figure 5-5. *The Discover tab on the Overview page of the deployed Speech service*

7. After trying the Speech-to-Text service demo, ensure that you click "Keys and Endpoint" to get the information we will use later when testing the API. Please record the key(s), endpoint, and region string somewhere convenient as we will use that information to connect to the service.

Now that you have the Speech service created, let's have a deeper look at the Speech-to-Text API and Text-to-Speech API.

What Is the Speech-to-Text API?

In recent years, search using speech has become increasingly important. People use smart assistants to manage their living and work environments. Those smart assistants are designed to accomplish tasks following the audio instruction provided by the user. In other words, it converts the audio input to a computer-readable command. For instances, when we search for an email, turn on and off our home lights, or switch the song on our music device using our voice. With these audio instructions, Speech-to-Text makes the audio input integrated with the target system seamlessly.

Even before COVID-19, some schools were showing a tilt toward online learning. However, since COVID-19 began, students have spent massive amounts of time at home and learning through video or streaming audio rather than in-person instruction. Given that the instructors are using that technology to deliver their material, the students may want to take notes and summarize the content via a Speech-to-Text API. If the trend continues and students begin taking advantage of this in online classes, the next question becomes, how can they manage all of this text content derived from the teacher's vocal instruction?

That is where the next level of interacting with this content comes in. This data can be indexed, searched, and managed all via this API. Is that applicable for a single student? Maybe not. Is it applicable to an application developer developing an app for students to do just that? Definitely.

Beyond these educational scenarios that may be relevant to many, this API's ability to transcribe, index, manage, and search this data applies to anyone who has to work with audio content. Perhaps the customer service department at your company, your friend who produces a podcast, or a professional colleague who works for an online healthcare provider could make use of these abilities to get better at their job or even leverage their newfound technical abilities to transition into technical roles.

What Is the Text-to-Speech API?

Text-to-speech services are often referred to as a TTS service or artificial speech synthesis. As you might imagine, people use these services to convert documents, web content, and blog posts into readily accessible audio. Beyond that, technology interfaces (like the smart assistants referenced in the preceding) make use of the ability to easily convert text to speech.

With the Text-to-Speech API, you can generate human-like audio from written text. At the time of publishing, it provides both standard and neural voices in more than 45 languages and locales and over 75 nonneural standard voices.

Using its asynchronous synthesis of long audio, the service gives you the ability to work with long sections of text as it will do the heavy lifting of parallelizing it and synthesizing the audio result. The standard voices and Speech Synthesis Markup Language (SSML) can make the artificial voice sound more natural and clearer to your end users. The user can also adjust its pitch and volume, add pauses, improve its pronunciation, and modify its speaking speed.

If you would like to experiment with the service beyond the basic examples provided by Microsoft, we recommend familiarizing yourself with Speech Synthesis Markup Language to customize speech characteristics, as well as focusing on the use of the neural voices.

Samples

There are five ways to interact with the Speech-to-Text API. If you are looking for the most detailed level of the API document, it is recommended to check the Swagger documentation for the v3 API. If you want to pick it up quickly with Python scripts, search specifically for the Speech SDK – Python session.

Swagger

If you are not familiar with it, Swagger is an Interface Description Language for describing RESTful APIs expressed using JSON. It is open source software with automated documentation, code generation, and test case generation functionality.

When you work with an API service with Swagger documentation, it usually provides a web page that shows all the endpoints that it has. For each endpoint, the Swagger document shows how you can interact with the API using POST, DELETE, GET, or PATCH to interact with the endpoint. It can also provide guidance on which data model you should send as a request object.

To visit the Swagger document, use this URL: `https://westus.dev.cognitive.microsoft.com/docs/services/speech-to-text-api-v3-0`. The "westus" can be replaced by your selected region string. It will auto-populate when selecting the target region for testing.

When visiting the Swagger document, the v3 API information of the Speech service should look similar to Figure 5-6. Please note that it also contains the Batch Transcription and Customization speech features mentioned earlier as well. For each endpoint, you can send a request through this page using the authentication key we recommended you store in a convenient place earlier in the chapter. Hopefully you did so – if not, click back into your Speech service and copy those keys. Besides Postman (which many developers reading this book are likely familiar with), Swagger provides you an alternative way to test this API and others.

Figure 5-6. *The Swagger document of the deployed Speech service*

Speech SDK – Python

As mentioned previously, the Speech SDK supports the Python package and aims to give Python developers a solid experience when interacting with the various Speech APIs.

In this part of the chapter, we will use the Python SDK to interact with the Speech-to-Text API. The pypi page of the Python SDK, azure-cognitiveservices-speech, is located here: https://pypi.org/project/azure-cognitiveservices-speech/. The Python package management used for this sample is Pipenv. It helps you manage your Python version and package environment more easily than in some other ways. Check out the Pipenv site if you are unfamiliar with it and would like to learn more about it: https://pipenv.pypa.io/en/latest/.

It is also good to note that this example can be reproduced on Mac, Windows, or Linux. No matter your computing platform of choice, you can follow along with this example.

In the Chapter 5 sample code folder, you will see a file called Pipfile. That file describes what Python packages along with the package version you need to run this project. By running *pipenv install*, the Python dependency will be installed in a virtual environment. Then, after running *pipenv shell*, you will be able to use this virtual environment with the dependency in the Pipfile to run the script.

Our first sample will cover using the Speech-to-Text API with input from a microphone. When interacting with the service, the speech key and the service region are the most important information you will need to initiate the speech recognizer. As shown in Listing 5-1, you should always start by setting up the resource to connect with the Azure API. With this setup, you will be able to recognize the audio content from your microphone.

Also, similar to the environment variable storage discussed in Chapter 4, it is recommended to store items like API keys in environment variables for greater security when uploading your code to locations like GitHub. Without that, your keys will be exposed to prying eyes. For test purposes and eliminating silly mistakes early on, it is alright to skip that step.

Listing 5-1. Initializing the Speech Recognizer for Microphone Input

```
speech_config = speechsdk.SpeechConfig(speech_key, service_region)
speech_recognizer = speechsdk.SpeechRecognizer(speech_config=speech_config)
result = speech_recognizer.recognize_once()
```

Note This code block displays the most important part for the script. Additional logging, error handling, etc. can be added as your project matures.

Our second sample involves using Speech-to-Text with static file input. To call the service using a static file, prepare the speech configuration similar to the previous sample, but in this case, we will be using an audio file (.wav). You can store the file as a local file or upload it to cloud-based storage such as Azure Blob Storage. Next, initiate the speech synthesizer with the audio configuration and the language. The audio configuration contains the settings of the input file, which can be a static file (as in this case) or a streaming data input (as you continue to experiment). After configuring these settings, you are ready to run this script and get the transcript from the audio. Listing 5-2 shows you how to initialize the speech synthesizer for static file input.

Besides the initiation of the synthesizer and the recognizer, there should be a block of code to handle the result status and result failure details. It is also recommended to log any messages when the result is received. This will help us have an easier way of debugging the interaction between our service and the Speech API.

Listing 5-2. Initializing the Speech Synthesizer for Static File Input

```
file_config = speechsdk.audio.AudioOutputConfig(filename=file_name)
speech_synthesizer = speechsdk.SpeechSynthesizer(speech_config=speech_
config, audio_config=file_config)
result = speech_synthesizer.speak_text_async(text).get()
```

Our third and final example in this section uses the Text-to-Speech service. This is my favorite as it can bring the beauty of text and literature to people who may struggle with their version (and whose struggles affect their ability to experience these texts). Here, we prepare the settings with the Speech API key similar to the previous samples. We also initiate the speech synthesizer with the speech configuration and the audio configuration. As you become more advanced in your skills here, you can also select different speaking voices so that you may experience the text with different voices as well. The basics are depicted in Listing 5-3. I hope you enjoy this one!

Listing 5-3. Sample of Text-to-Speech

```
speech_config = speechsdk.SpeechConfig(subscription=speech_key,
region=service_region)
speech_synthesizer = speechsdk.SpeechSynthesizer(speech_config=speech_
config)
text = "The first text to speech sample."
result = speech_synthesizer.speak_text_async(text).get()
```

Thoughts   For the Text-to-Speech service, there are several voice candidates you may choose as the output audio. After trying each of them, how do you feel about the different voices? How do the different voices affect the user experience?

Register Your Voice

Besides the Speech-to-Text and Text-to-Speech services, Azure also provides a service called Speaker Recognition. While Speaker Recognition is still in preview as of publishing time, it is an important part of the overall Speech service and worth our time to explore. So what is this service and why is it important? Let's examine that.

Speaker Recognition is a service that can identify who is speaking based on the characteristics of their voice. Audio can reveal a huge amount of information, including physiological and behavioral characteristics, and Speaker Recognition leverages all of these abilities to help identify and verify speakers.

While there are a wide variety of use cases to explore with this service, its main uses currently are for things like security and forensics. This service, by empowering you to identify and verify a speaker, can be used to secure a facility or even identify someone at the scene of a crime or security incident. This is why, even though the service is referred to as Speaker Recognition, it is commonly referred to as speaker identification and/ or speaker verification. Let's discuss, at a high level, both of these components of the service.

Speaker Verification

Speaker verification does exactly what its name would indicate – it uses one's voice to verify that the identified person is the speaker. Typically, the system has stored the user's voice signature so that it is preregistered in the system. When a new stream of audio comes in, the system parses that stream for the particular features and characteristics of the audio.

This verification can happen in a couple of different ways – a text-dependent method and a text-independent method. When the signature and the incoming audio use the same text as the script that has been read, that is referred to as the text-dependent method. When the voice signature is not using the same words as the incoming audio, that is referred to as the text-independent method. Regardless of method, the verification happens largely the same way. The system compares the signature and the incoming audio and, if the differences lie within a certain threshold, the system will say this voice belongs to the person. If the differences are above the threshold value, it will say otherwise.

Whether a business is using this for customer identity verification in call centers or to facilitate contactless facility access or something else entirely, you can imagine the possibilities of touchless/imageless verification for a variety of businesses and applications.

Speaker Identification

Speaker identification isn't quite as straightforward as its given name might suggest. It is used to identify the different speakers in an audio recording or the audio track of a video recording. You might think of it as speaker differentiation rather than speaker identification, but that is not how Microsoft has named it, so that is not how we will refer to it. Similar to verification, however, generally speakers will register their voices prior to the event, that registration will be stored, and then speaker identification will try to match the audio in the meeting record to those registered voices. Currently, Speaker Recognition is capable of recognizing up to 50 speakers in each request.

Figure 5-7 lays out the difference between these two capabilities of Speaker Recognition visually. In speaker identification, the number of decision alternatives is equal to the size of the observation group. In verification, there are only two choices, acceptance or rejection, as there are multiple voices for the service to sort through.

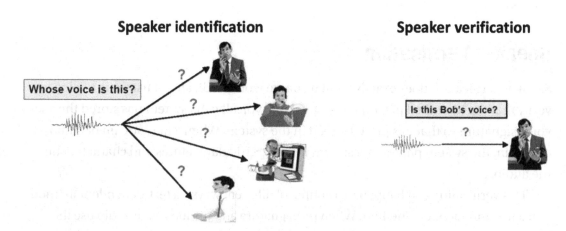

Figure 5-7. *The differences between speaker identification and speaker verification*

Security and Privacy

Obviously anytime we are talking about personal data, especially personal data interacting with or being stored in the cloud, security and privacy of that data are foremost concerns. As your apps and projects mature from personal projects to professional projects, it is important to consider these issues as you decide where to store the speech data we have discussed in the preceding. While there are a wide variety of options, we have recommended Azure Blob Storage in the preceding due to the robust security inherent in Azure data centers as well as the detailed methods it has to lock access down to very specific users and roles within your Azure subscription. Whatever storage solution you choose to use, please keep these things in mind for the safety and security of your users.

Aside from storage, however, the Speech service itself has a role to play in properly securing this data. Within the service, you can specify how long it will be stored along with specifying when to update and/or delete a particular speaker's data when the audio is registered with the API. Lastly, when the service has been deleted from your subscription, it is good to remember that the related stored data will be deleted as well. Now that we have walked through some of its capabilities, let's look a bit deeper into how to use this API.

How to Use the Speaker Recognition API

Here we will again use the Speech SDK referenced earlier, this time to create a speaker profile. We will then use that speaker profile to identify the speaker.

Note   At publishing time, Speaker Recognition is currently only supported in Azure Speech resources created in the westus region.

As we begin, here are a few prerequisites for you to keep in mind. First, you will need to create a Speech service in the westus region (see the preceding note). Second, you will need to create an Azure Blob Storage account and upload a spoken word audio file to it. Third, get the subscription key from the Speech service created earlier. For reference, the location of the key is as shown in Figure 5-8.

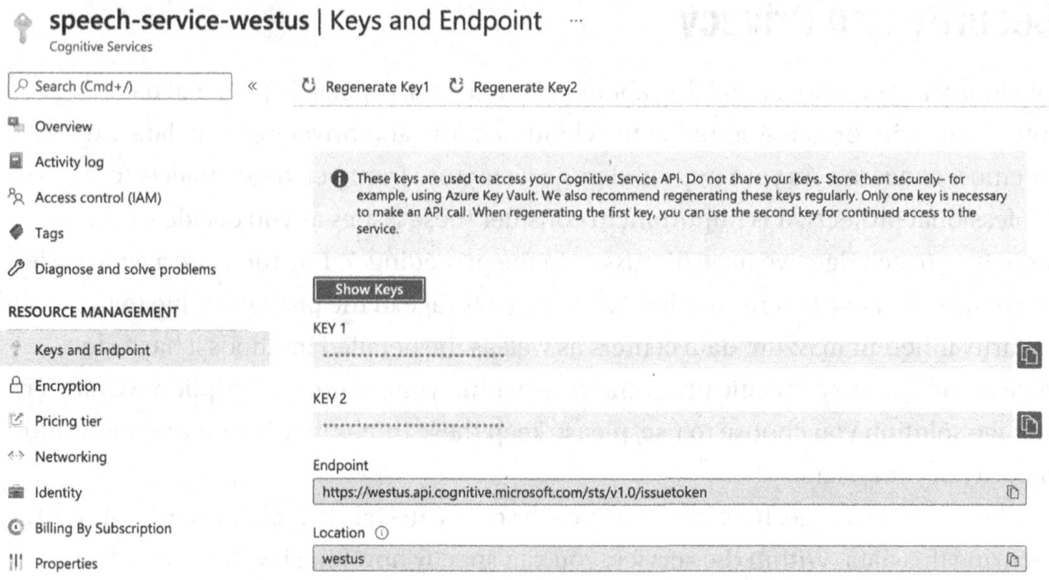

Figure 5-8. *The key information page of the deployed Speech service*

Our next example will dive into the speaker registration described in the preceding. In this sample, we will interact with the Speech service with a script released from Microsoft that is not directly a part of the Speech SDK as of publishing time. There are two main concepts in this example: create a profile for the audio and then enroll/register that profile. Following that, we will print out all the existing profiles to ensure that you have correctly registered the audio.

As you would expect, you will need the subscription key for your Speech service as you have needed it for every other example. Go ahead and retrieve that from its safe and convenient place and then use Listing 5-4 to get the enrollment response containing the profile ID and file path.

Listing 5-4. Speaker Registration

```
helper = IdentificationServiceHttpClientHelper.IdentificationServiceHttpCli
entHelper(subscription_key)
creation_response = helper.create_profile(locale)
profile_id = creation_response.get_profile_id()
```

```
enrollment_response = helper.enroll_profile(profile_id, file_path, force_
short_audio.lower() == "true")
profiles = helper.get_all_profiles()
```

The final example in this section will cover speaker identification. After listing all the existing speaker profiles in the previous example, now you may test speaker identification with the prepared recorded file. To identify the speaker, you will need the target audio file path and profile IDs. Keep in mind that, for future experimentation, you may give more than one profile ID to be identified. The Python to perform this identification is shown in Listing 5-5.

Listing 5-5. Speaker Identification

```
helper = IdentificationServiceHttpClientHelper.IdentificationServiceHttpCli
entHelper(subscription_key)
identification_response = helper.identify_file(file_path, profile_
ids,force_short_audio.lower() == "true")
```

The output JSON for the speaker identification should look like the image in Figure 5-9. The maximum number for speaker identifications is 2, and you have to make sure the diarizationEnabled flag is set to true as well.

```
"speaker": {
    "$id": "#root/recognizedPhrases/items/nBest/items/speaker",
    "title": "Speaker",
    "type": "integer",
    "description": "if `diarizationEnabled` is `true`, this is
    "examples": [
        1
    ],
    "default": 1,
    "minimum": 1
},
```

Figure 5-9. *The screenshot of the speaker identification output JSON*

> **Tips**   If you are looking for C# .NET samples, visit `https://github.com/Azure-Samples/cognitive-services-speech-sdk/tree/master/quickstart/csharp/dotnet.`

Speech Translation SDK

The Speech Translation SDK provides us the ability to work with the Speech Translation endpoint. The Speech Translation endpoint can translate the audio to more than 30 languages as well as customize the translation with terms you provide. In Figure 5-10, I have categorized the modules into three main components by function (utilities, main interface, and implementation).

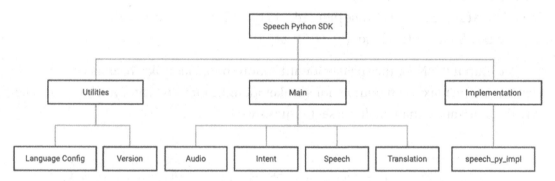

Figure 5-10.   *The module structure of the Speech Translation Python SDK*

Let us walk through the structure of the Speech Translation SDK. First, we'll discuss the utilities. These modules take care of the initial settings and lay the groundwork for the rest of our work. The LanguageConfig module handles the language configuration for a client. The AutoDetectSourceLanguageConfig helps the setup for the specific endpoint as well as the list of potential source languages to detect (that contains the SourceLanguageConfig). The SourceLanguageConfig is then used to help the setup for the source language and customized endpoint.

Version is much simpler – it does as you would expect and shows the current version number.

The main functions are a bit more in depth as these are the modules that interact with audio data and the Speech API interface with the predefined LanguageConfig.

The first main function is Audio. It handles the input and output for the audio data with configurations that set up the audio input and output. If the data source is a streaming one, it also provides pull and push settings to govern the streaming audio.

Next, we have the Intent function. As you might guess, the IntentRecognizer originates from this module. It is the most important component when implementing the intent to identify tasks. From this, the LanguageUnderstandingModel can be initiated with the subscription information, endpoint, or application ID. If everything works as planned, the IntentRecognitionResult will provide the output back to you. If things do not go as planned, the reason will be presented in the CanceledEventArgs class.

Moving on from the Intent function, we have the Speech function. As discussed previously, this module interacts with both the Text-to-Speech and Speech-to-Text services. We have covered these in some depth so far, so let us move on to the Translation function.

The Translation function contains the SpeechTranslationConfig that can be initiated with the subscription information, endpoint, host, or an authentication token. The TranslationRecognizer also includes the synthesizer in it, which takes care of the translation result aggregation as well.

Finally, we will discuss the implementation functions. The speech_py_impl module contains the implementation of all the classes, including (but not limited to) Activity, Audio, Speech, BotFrameworkConfig, Conversation, Dialog, Intent, Translation, Connection settings, and the Session event.

Tips   When you develop the service with the samples we provide, try to start with the minor changes. Understand the key elements to make these scripts work, work through the flow of the REST API, and supplement this work with the official documents to look for what the mandatory and optional parameters are and what abilities those may provide. It will also help you get more familiar with the SDK and build deeper knowledge of these services.

Now that we have introduced the interface, you have most of the knowledge you need to start using this package. For your reference, the prerequisites of different operating systems to install the package are as follows:

1. *Windows* – x64 and x86

2. *Mac* – macOS X version 10.12 or later

3. *Linux* – Ubuntu 16.04/18.04, Debian 9, RHEL 7/8, CentOS 7/8 on x64

Use the pip install to install the Cognitive Services Speech SDK Python package:

```
pip install azure-cognitiveservices-speech
```

Before we get to the next example, it is important to have your environment fully set up. In order to have an environment set up to proceed with the rest of the examples, please ensure that you have installed Visual Studio Code, installed Python 3.8, and used Git to pull the sample repository for this book. Once that is complete, run `pipenv install` and `pipenv shell` to initiate the development environment within Visual Studio Code. Finally, ensure that the Speech API service discussed previously has been created and is still present in your Azure account.

Translate Speech to Text

For the first example in this section, we will again be using the Speech service with the Python SDK and create a speech recognizer object with our subscription. As before (you're likely getting used to this by now!), copy your subscription key and region information from your Speech service as shown in Figure 5-11.

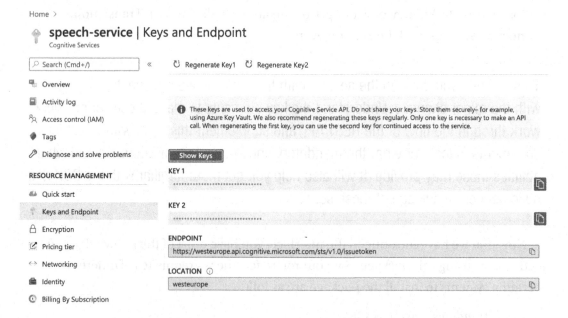

Figure 5-11. *The key information in the Keys and Endpoint page*

When initiating the SpeechRecognizer, you will need the SpeechConfig. To create a SpeechConfig (which contains the subscription key and region information noted in the preceding), you can combine that information with the sample from Listing 5-6. Using the SpeechRecognizer object, you can start the recognition process for a single utterance.

Listing 5-6. Initiate Speech Recognizer

```
speech_config = speechsdk.SpeechConfig(speech_key, service_region)
speech_recognizer = speechsdk.SpeechRecognizer(speech_config=speech_config)
result = speech_recognizer.recognize_once()
```

Next, inspect the TranslationRecognitionResult returned from the code in Listing 5-6. As Listing 5-7 shows, for different types of recognition results, we can use the different log messages here. When the result is returned, you will be able to see if the audio has been parsed without error. You can also identify the error type from the error handling included.

Listing 5-7. Speech Recognizer Result Processing

```
if result.reason == speechsdk.ResultReason.RecognizedSpeech:
    print(f"Recognized: {result.text}")

elif result.reason == speechsdk.ResultReason.NoMatch:
    detail = result.no_match_details
    print(f"No speech could be recognized: {detail}")

elif result.reason == speechsdk.ResultReason.Canceled:
    cancellation_details = result.cancellation_details
    reason = cancellation_details.reason
    print("Speech Recognition canceled: {reason}")

    if cancellation_details.reason == speechsdk.CancellationReason.Error:
        detail = cancellation_details.error_details
        print(f"Error details: {detail}")
```

Translate Speech to Multiple–Target Language Text

If you have audio translation functionality in your application, you will need to convert the audio content to a multilanguage transcript.

First, as expected by now, copy your subscription key and region information from your previously created Speech service. This is shown in Figure 5-12.

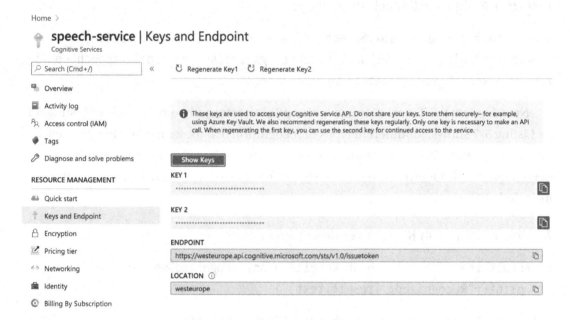

Figure 5-12. *The key information in the Keys and Endpoint page*

Next, create a SpeechTranslationConfig with the subscription key and region information. Update the SpeechTranslationConfig object with the speech recognition source language. Then, update the SpeechTranslationConfig object with your specified translation target language. As Listing 5-8 shows, the target language can be more than one language as well.

Listing 5-8. Add attributes to Speech Config

```
translation_config = speechsdk.translation.SpeechTranslationConfig(subscrip
tion=speech_key, region=service_region)
fromLanguage = "en-US"
translation_config.speech_recognition_language = fromLanguage
translation_config.add_target_language("de")
translation_config.add_target_language("fr")
```

176

Next, as shown in Listing 5-9, create a TranslationRecognizer object using the SpeechTranslationConfig object from the preceding. Using the TranslationRecognizer object, start the recognition process for a single utterance.

Listing 5-9. Build the Translation Config

```
recognizer = speechsdk.translation.TranslationRecognizer(translation_
config=translation_config)
result = recognizer.recognize_once()
```

Following that, inspect the TranslationRecognitionResult returned. As Listing 5-10 shows, we will process the result from the Cognitive Service with a variety of reasons. We will then print out the multilanguage translated text in the application.

Listing 5-10. Translation Result Processing

```
if result.reason == speechsdk.ResultReason.TranslatedSpeech:
    print(f"RECOGNIZED {fromLanguage}: {result.text}")
    de_translate = result.translations["de"]
    fr_translate = result.translations["fr"]
    print(f"TRANSLATED into de: {de_translate}")
    print(f"TRANSLATED into fr: {fr_translate}")

elif result.reason == speechsdk.ResultReason.RecognizedSpeech:
    print(f"RECOGNIZED: {result.text} (text could not be translated)")

elif result.reason == speechsdk.ResultReason.NoMatch:
        detail = result.no_match_details
        print(f"NOMATCH: Speech could not be recognized: {detail}")

elif result.reason == speechsdk.ResultReason.Canceled:
        print("CANCELED: Reason={}".format(result.cancellation_details.
        reason))
        reason = result.cancellation_details.reason
        if reason == speechsdk.CancellationReason.Error:
            detail = result.cancellation_details.error_details
            print(f"CANCELED: ErrorDetails={detail}")
```

Translate Speech to Speech

Our final example in this section is using the Speech-to-Speech service, which translates one spoken audio input into spoken words in another language.

As usual, the first step is copying your subscription key and region information from your Speech service. For reference, that is shown in Figure 5-13.

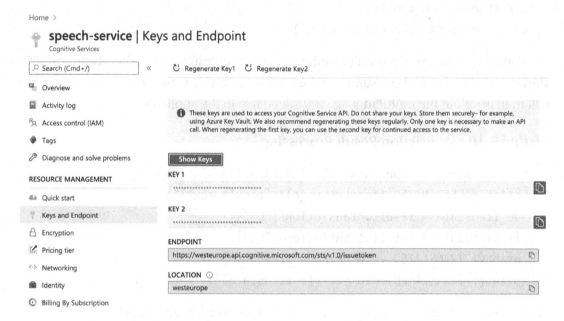

Figure 5-13. *The key information in the Keys and Endpoint page*

Next, create a SpeechTranslationConfig, which contains the subscription key and region information. Following that, update the SpeechTranslationConfig object to specify the speech recognition source language and the translation target language.

The next step is to choose the voice name from the standard voice list, which contains more than 75 standard voices that are available in over 45 languages and locales. For the most current list, please refer to https://docs.microsoft.com/en-us/azure/cognitive-services/speech-service/language-support#standard-voices. For example, for Chinese (Taiwanese Mandarin), the language code is "zh-TW", and there is a female voice name called "zh-TW-Yating". The name makes sense as that is the most common female name in Taiwan. While that is interesting, the voice chosen for this example is "de-DE-HeddaRUS".

Now let us initiate the recognizer with the translation config. The code for the previous two paragraphs is shown in Listing 5-11.

Listing 5-11. The Speech Translation Config

```
translation_config = speechsdk.translation.SpeechTranslationConfig(
        subscription=speech_key,
        region=service_region)

fromLanguage = "en-US"
toLanguage = "de"

translation_config.speech_recognition_language = fromLanguage
translation_config.add_target_language(toLanguage)
translation_config.voice_name = "de-DE-HeddaRUS"

recognizer = speechsdk.translation.TranslationRecognizer(
        translation_config=translation_config)
```

In Listing 5-12, we will set up the callback function and start recognizing the audio input.

Listing 5-12. Callback and Start Recognizing

```
def synthesis_callback(evt):
    size = len(evt.result.audio)
    status_msg = "(COMPLETED)" if size == 0 else ""
    print(f"AUDIO SYNTHESIZED: {size} byte(s) {status_msg}")

    recognizer.synthesizing.connect(synthesis_callback)

    print("Say something...")
    result = recognizer.recognize_once()
```

Next, we receive the response and process it according to the specified response reason. You can see the code for those things in Listing 5-13.

Listing 5-13. Process the Response

```
if result.reason == speechsdk.ResultReason.TranslatedSpeech:
    print(f"RECOGNIZED {fromLanguage}: {result.text}")
    de_translate = result.translations["de"]
    print(f"TRANSLATED into {toLanguage}: {de_translate}")

elif result.reason == speechsdk.ResultReason.RecognizedSpeech:
    print(f"RECOGNIZED: {result.text} (text could not be translated)")

elif result.reason == speechsdk.ResultReason.NoMatch:
    reason = result.no_match_details
    print(f"NOMATCH: Speech could not be recognized: {reason}")

elif result.reason == speechsdk.ResultReason.Canceled:
    reason = result.cancellation_details.reason
    print(f"CANCELED: Reason={reason}")

    if reason == speechsdk.CancellationReason.Error:
        detail = result.cancellation_details.error_details
        print(f"CANCELED: ErrorDetails={detail}")
```

Advanced Speech Knowledge

After all the examples in the previous section, the hope is that you have deeper understanding about the Cognitive Services Speech service and how to interact with it. If that has piqued your interest, you should enjoy this section where we delve more into how these abilities actually work. You will get some additional details here and, with that, hopefully a deeper understanding of the Speech service and a better knowledge of the domain terminology here as well.

How Do Speech-to-Text and Speech Recognition Work?

As you have learned throughout this chapter, there are a wide variety of scenarios where we can use the Speech service. Here, I select "speech sentiment analysis service" as a study case. The research paper is called "Sentiment Analysis on Speaker Specific Speech

Data" from the 2017 International Conference on Intelligent Computing and Control Systems. I have found this to be a great example to read and study and hope you find it the same way.

In past decades, people have used textual sentiment analysis to analyze message board text, news articles, and even restaurant reviews. The results of this analysis can bring us deeper insights about our customers, the sentiment of the general public, and overall market trends.

Using the Speech service, we can apply this same idea to audio data as well. We can analyze the audio of a recorded call to our company's customer service line, the identity and emotion of speakers in an audio or video meeting, and other scenarios as well. While audio sentiment analysis may not be quite as mature as textual sentiment analysis, we can still use the same concepts to understand our audio data better.

To work with the speech sentiment analysis service, we will need three components to achieve this. First, from the diagram in Figure 5-14, you can find there is "speaker discrimination." When we analyze the speech data, it does not always come with a single person record; it sometimes (likely often) comes with several people's voices in the same file. You will need to identify the audio resources to make for more consistent analysis. What is referred to in the following as speaker discrimination is also called Speaker Recognition.

You don't necessarily need someone's real name to label the data. As long as you are able to separate each person's speech in the recording, you are able to perform sufficient Speaker Recognition.

The second component we will need is "speech recognition" (seen on the bottom side of Figure 5-14). You're likely reading this after reading the earlier paragraph and thinking, "This is confusing terminology." It can be a bit, but the "speech recognition" here may be more accurately understood as "speech parsing," meaning converting the audio file to short text or phrases, which can then be understood by humans or computers. Because these models are quite complex, users do not often create their own model to do this. Previously, common tools for this included Bing Speech or Google Speech Recognition as the speech recognition service of choice. Cognitive Services aims to change that, of course.

If you follow along with Figure 5-14, you will see that we need two labels from the Speech service. One is speaker discrimination and the other is speech recognition. Once you have the speaker ID and speech transcription, you are then able to parse the audio data to a series of textual data.

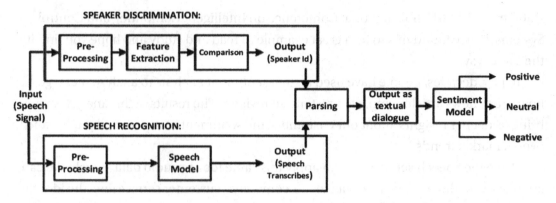

Figure 5-14. *Proposed structure for the sentiment analysis system from "Sentiment Analysis on Speaker Specific Speech Data"*

After we have recorded a series of conversational dialog, each sentence becomes an input for the sentiment model. In the end, we can see the trend of emotions from different speakers' conversational content. The data output will reveal who started a great discussion, as well as when people started losing focus or had emotions triggered by certain words or sentences during the discussion. We will have timestamps for all of this as well.

This service will be able to make the audio data more searchable. Thinking back to scenarios discussed in earlier chapters, audio, video, and podcast content can be searched, indexed, and managed in a much easier fashion using a service like this.

One thing we haven't talked about is how we can identify the speakers. Generally, we will identify the speakers by using adjectives to describe the different voices. Examples include hoarse, flat, deep, loud, throaty, smooth, taut, low, high, shrill, and many, many more. How does a computer disguise these different types of voices and separate them into different speaker profiles?

Usually people use "Hertz" as the first data point in this analysis. Most men typically range between 85 and 180 Hz and most women between 165 and 255 Hz. While this is a basic data point to use, it is not only the range of the Hertz that is significant. Different languages also sound differently, because people use different ways to pronounce the words and speak to people. In addition, vocabulary and expressions vary widely from language to language as you would expect.

There is an interesting article on this topic written by Erik Bernhardsson. He presented a table compiling information with the gender of the speaker and the spoken language, mapping the value of the FFT (Fast Fourier Transform) coefficient magnitude [2]. The table in Figure 5-15 shows different peak values of female speakers across different languages. Also, the different patterns of languages really reflect how different languages do indeed sound very different.

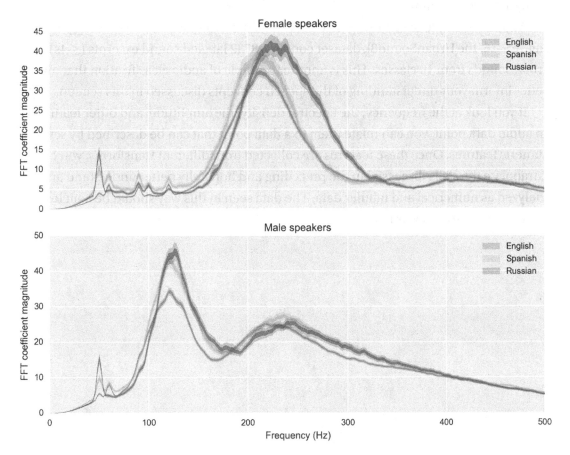

Figure 5-15. *Comparing languages by language pitch from reference [2]*

Besides FFT, you will see a variety of types of Fourier Transforms when reading the audio analysis topic article. The mathematic meaning of the Fourier Transform is beyond the scope of this book, but, if you are interested in it, the article identified in the "References" section as [2] is well worth a read. The key concept is that the Fourier Transform is a transform that can help you transfer the audio data to a series of numbers that can convert a continuous signal from time domain to frequency domain. Once this conversion has taken place, the data is much easier to analyze.

If you visualize the audio data, you may find there are shared patterns when different people pronounce the same word in the same language. That is the reason why people can use similar metrology to classify audio data with a neural network as they do in the Computer Vision realm.

Another example of this is the visualization of the UrbanSound8k dataset. [1] Seen in Figure 5-16, the UrbanSound8k dataset contains 8732 labeled sound excerpts ($\leq$ 4s) of urban sounds from 10 classes. This is another example of audio classification that may better inform your understanding of the speech concepts discussed in this section.

If you look at the frequency, the spectral density, the amplitude, and other features of an audio data point, you can relate them to a data point that can be described by several numeric features. Once these features are collected from different transforms, you can extrapolate the skills discussed in the preceding and hopefully better understand audio analyzed as numerical and tabular data. The data seen in this way should be a bit less abstract to you than it was before.

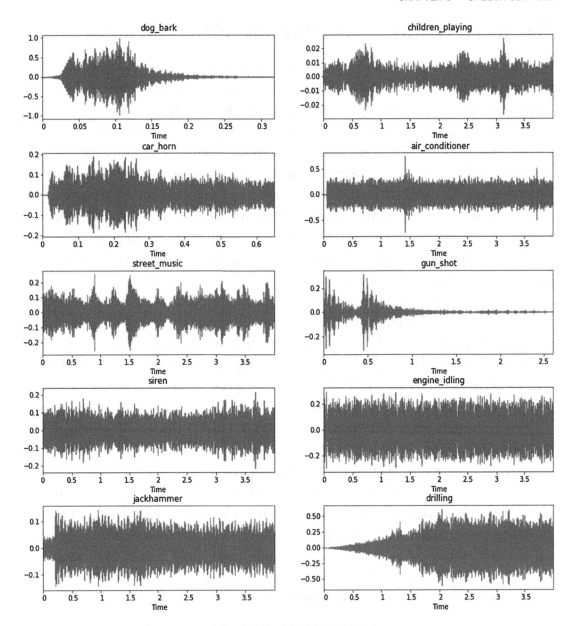

Figure 5-16. *Visualization of the URBANSOUND8K dataset*

The Speech FSDD Dataset (Advanced Speech Case)

Now that you better understand the mechanics of how Text-to-Speech and Speech-to-Text work, you may wish to try out the MNIST (Modified National Institute of Standards and Technology) dataset of speech recognition, Speech FSDD (Free Spoken Digit Dataset).

The MNIST dataset is often the first dataset used by people who are beginning to study deep learning and, most especially, Computer Vision (as we've discussed elsewhere in this book). It was made from 30,000 patterns of handwriting digital images. People can use the dataset to learn how to build a basic image classification model.

Because we're speaking about speech recognition here, though, let's begin with a similar "hello world" level of audio dataset. FSDD is an open dataset for people learning how to perform speech recognition. In order to "conquer" this dataset, you will need to identify spoken digits from 1,500 recordings.

There are three speakers recorded in this dataset. Each of them recorded each digit 50 times. The size of this dataset is 10 MB.

Now that the stage is set, let us learn more about how to work with the audio data of FSDD. As in the rest of the section, Python is the language of choice here. However, if you are a C# developer, you can refer to [3] in the "References" section to read more about how to use C# to interact with this dataset.

First, clone the sample code from our sample GitHub repository, and download the dataset from `https://github.com/Jakobovski/free-spoken-digit-dataset`. Download it to the folder where you will work with it later.

Next, import the libraries detailed in Listing 5-14 for later usage. The package librosa is the main audio analysis package we use here. It will extract the audio features for you. In addition, particularly for data people who are reading this book, there are some packages with which you might be familiar: NumPy and Pandas as basic ELT packages for Python. The matplotlib and librosa.display packages take care of the visualization for the audio features. Finally, Keras and sklearn help with building the machine learning models.

Listing 5-14. Import the Libraries

```
import numpy as np
import matplotlib.pyplot as plt
import librosa
import os
```

```
import librosa.display
from keras.layers import Dense
from keras.models import Sequential
from keras.optimizers import SGD
import pandas as pd
from sklearn.model_selection import train_test_split
```

Following the import, we will read the dataset and extract the STFT (Short-Time Fourier Transform) feature from the audio. STFT is a frequency feature representation for audio. After getting the audio feature, make a plot for each digit. The steps to do this are found in Listing 5-15.

Listing 5-15. Read the File, and Visualize It

```
file = os.listdir("free-spoken-digit-dataset/recordings")
data = []
for i in file:
    x, sr = librosa.load("free-spoken-digit-dataset/recordings/" + i)
    data.append(x)

def plot_images(images, cls_true, cls_pred=None):
    assert len(images) == len(cls_true) == 9

    # Create figure with 3x3 sub-plots.
    fig, axes = plt.subplots(3, 3, figsize=(15, 15))
    fig.subplots_adjust(hspace=0.3, wspace=0.3)

    max = np.max(images)
    min = np.min(images)

    for i, ax in enumerate(axes.flat):
        # Plot image.
        librosa.display.specshow(images[i], ax=ax, vmin=min, vmax=max)

        # Show true and predicted classes.
        if cls_pred is None:
            xlabel = "True: {0}".format(cls_true[i])
        else:
            xlabel = "True: {0}, Pred: {1}".format(cls_true[i], cls_pred[i])
```

```
    ax.set_xlabel(xlabel)
# Ensure the plot is shown correctly with multiple plots
# in a single Notebook cell.
plt.show()
```

After that work, we aim to plot a 3 × 3 graph. Each of them shows one of the digit files. From the labels in Figure 5-17, you will be able to see there are different feature values across different digits in the audio data. You can also see in that figure that many of the graphs have a similar color distribution. To compare to the other numbers, we need to separate them into ten categories. To do that, though, we will need to use a neural network classification.

Figure 5-17. *Visualization of the audio feature*

As seen in Listing 5-16, we then will split the train and test data. We will prepare the data using the STFT feature transformation and then put the audio data into a NumPy array. That can then be used later in the Keras neural network. Lastly, we'll use *get_dummies* to get the one-hot encoding for the data label.

Listing 5-16. Split X(feature) and Y(label) and Split Training and Testing Data

```
# prepare X and Y
X = []
for i in range(len(data)):
    X.append(abs(librosa.stft(data[i]).mean(axis=1).T))
X = np.array(X)
y = [i[0] for i in file]
Y = pd.get_dummies(y)  # One Hot Encoding

# Splitting Dataset
X_train, X_test, y_train, y_test = train_test_split(X, Y, test_size=0.25)
```

Next, we will create a small sequential neural network model with an input dimension = 1025. We will also add four Dense layers with SGD (stochastic gradient descent) as our optimizer. Listing 5-17 shows this code.

Listing 5-17. Create the Model and Start Training

```
# model:
model = Sequential()
model.add(Dense(256, activation="tanh", input_dim=1025))
model.add(Dense(128, activation="tanh"))
model.add(Dense(128, activation="tanh"))
model.add(Dense(10, activation="softmax"))
sgd = SGD(lr=0.01, decay=1e-6, momentum=0.9, nesterov=True)
model.compile(loss="binary_crossentropy", optimizer=sgd,
metrics=["accuracy"])
history = model.fit(
    X_train,
    y_train,
    epochs=20,
    batch_size=128,
```

```
    verbose=1,
    validation_data=(X_test, y_test),
    shuffle=True,
)
score = model.evaluate(X_test, y_test, batch_size=128)
```

For a deeper dive, you can also try this with different layers like dropout, conv, and others. Different activation functions like relu, tanh, sigmoid, and softmax can also be used. Trying out different optimizers is encouraged for broader learning as well.

Besides the deeper dive recommendations, though, you can see that after running the 20 epochs of model.fit(), you will get a model that can identify the digits with accuracy of 91%.

If you want to see the plot of the loss and accuracy during training, use what is found in Listing 5-18. You will see the accuracy graph that is outputted in Figure 5-18.

Listing 5-18. Create a graph for loss and evaluation loss value with different training epochs

```
plt.plot(history.history["loss"])
plt.plot(history.history["val_loss"])
plt.xlabel("Epoch")
plt.ylabel("loss")

# plot acc
plt.plot(history.history["loss"])
plt.plot(history.history["val_loss"])
plt.xlabel("Epoch")
plt.ylabel("loss")
```

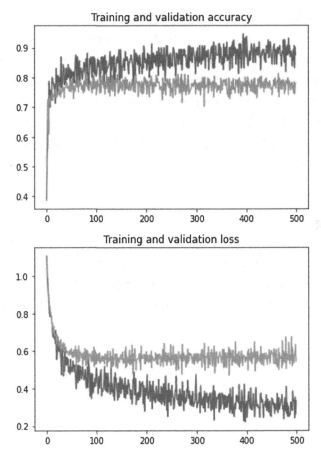

Figure 5-18. *Training accuracy and loss graph*

Summary

We began this chapter talking about some real-world use cases for the Speech service to encourage your own thinking about real-world scenarios where you could apply this technology. As we continued on from there into setting up the Speech service, using Speech-to-Text and Text-to-Speech services, and finally working through the audio and text data, it is our hope that none of this seems nearly as intimidating as it did before reading this chapter. Once you were more familiar with the Speech service, we then walked through Speaker Recognition so you could learn how to identify who is speaking in conversations and recordings. Then, with the Translation SDK, we showed how you can build the bones of an application that can communicate between two languages.

Finally, for those interested in a deeper dive into the concepts and mathematics behind these services, we introduced you to datasets and open source libraries that can help facilitate that deeper learning. You now also know how to visualize the audio data, and you understand which Python library to use to work with audio analysis.

There is certainly a mountain of information to learn to become an expert in everything we discussed in this chapter. That said, we hope you now have what you need to understand the capabilities of Cognitive Services Speech and so you can start your own Speech service project.

References

[1] https://urbansounddataset.weebly.com/urbansound8k.html

[2] https://erikbern.com/2017/02/01/language-pitch.html

[3] http://accord-framework.net/docs/html/T_Accord_Audition_BagOfAudioWords.htm

CHAPTER 6

Power Platform and Cognitive Services

Azure Cognitive Services is becoming more incorporated into applications in the Microsoft Ecosystem, including Microsoft Power Platform. For the business analyst, or developer just getting started with AI, this chapter will provide an overview of AI functionality available within the Power Platform that you can incorporate into your own solutions or use cases.

Power Platform is a collection of three products: Power BI to analyze and gain insights about your data, PowerApps to create apps on top of your data and Power Automate to build business processes, workflows or develop app logic. All of the three Power Products are built to provide developers with a no-code, low-code like experience, leveraging the basic building blocks and connectors that already exist on the platform.

The typical use cases for AI in Power Platform include but are not limited to applying predefined Cognitive Services models for sentiment analysis, detecting key phrases from sentences, detecting language from data, and tagging images for data. These abilities will enable you to enrich business intelligence (BI), apps, or automations with AI and speed up development processes, to drive better business outcomes.

The focus of this chapter is on the range of AI opportunities in Power BI, and by the end of the chapter, you will learn to use AI Builder, a Power Platform function that lets you incorporate AI into your apps and automation processes.

© Alicia Moniz, Matt Gordon, Ida Bergum, Mia Chang, Ginger Grant 2021
A. Moniz et al., *Beginning Azure Cognitive Services*, https://doi.org/10.1007/978-1-4842-7176-6_6

AI in Your BI Overview

Artificial Intelligence (AI)–Major vendors have invested heavily in
Artificial Intelligend (AI) driven applications over the past few years. Blending AI
and data analytics enhances organizational decision making. It enables the ability
to perform complex tasks and gain insights on large datasets. Power BI is a market-
leading enterprise and self-service business intelligence and analytics platform that
lets you collect and transform your data into rich visualizations so you can focus on
what matters for you. BI refers to a set of processes, architectures, and technologies
that converts data into meaningful information to drive business-critical decisions.
The ability to use prebuilt AI functionality within products like Power BI opens new
possibilities, even for those who do not know how to write a single line of code.
Analysts can perform sentiment analysis from social media, customer service, or
online comments and extract key phrases from text to visualize the feedback people
are providing in regards to their services or brands. Further options are available to
extend AI for the data scientist who can incorporate Azure Machine Learning models
using languages such as R or Python. These options allow data professionals to work
alongside the business analyst. It becomes an ecosystem where people of different skill
levels can collaborate to incorporate cognitive services using different options. Let's
take a look at some different options where AI can be incorporated across the Power
Platform. Table 6-1 summarizes some of the key options for two types of personas, data
scientist and business analyst/low-code developer across Power Platform.

Table 6-1. *Summarization of AI Options for Personas on Power Platform*

Persona	Options	Product(s)	It lets me…	Select when…
I am a **business analyst or low-code** developer without a strong background in machine learning.	Include Machine Learning into your solution with **Automated Machine Learning (AutoML)** without having an Azure subscription.	Power BI Service – Dataflows and Premium	Apply, train, and see results of a Machine Learning model to perform binary prediction, general classification, and regression. **This will not be covered in this chapter.**	You have Power BI Premium, simple machine learning use cases, and no Azure access.
	Include Cognitive Services into your solution with p**rebuilt Cognitive Services** and AI Insights without having an Azure subscription.	Power BI Service – Dataflows, Desktop, and Premium	Access and consume models for sentiment analysis, key phrase extraction, language detection, and image tagging to enrich your data.	You have Power BI Premium EM2, A2, or P1 and a generic use case for data enrichment.
	Integrate Power BI with **Cognitive Services APIs** in Azure and create custom functions in Power Query using the M language.	Azure Cognitive Services (Azure subscription) and Power BI Desktop or Power BI Online – Dataflows	Access and consume models for Anomaly Detection to visualize anomalies throughout your time-series data or Text Analytics for sentiment analysis, language detection, or key phrase extraction.	You have Azure Cognitive Services that you would like to integrate with a Power BI solution.

(*continued*)

195

Table 6-1. (*continued*)

Persona	Options	Product(s)	It lets me...	Select when...
	Include **prebuilt AI visuals** into your solution to analyze and interpret data.	Power BI Desktop or Power BI Service	Use the key influencers visual to identify drivers for increase or decrease of KPIs, decomposition tree visual to visualize data across multiple dimensions and drill down based on high or low contribution to aggregation, and Q&A visual for natural language querying and suggested questions based on underlying data.	You want to add intelligence into your visuals in a Power BI solution.

(*continued*)

Table 6-1. (*continued*)

Persona	Options	Product(s)	It lets me...	Select when...
	Include **AI Builder** custom model types like prediction, form processing, object detection, category classification, and entity extraction and integrate into your PowerApps or Power Automate solutions. Or use the **AI models** already built and trained and published to do a specific task, like reading content of a business card, recognizing text, sentiment analysis, language detection, etc.	PowerApps or Power Automate and AI Builder	Use AI Builder to add intelligence into your apps or automation scenarios. You can improve business performance by automating processes or predicting outcomes – either purpose built for unique requirements or prebuilt for generic use cases.	You have an AI Builder license. You have a generic or custom AI use case either including an app for data entry or data output or a business process that you would like to automate.
I am a **data scientist or developer** with a background in machine learning.	Include Machine Learning into your solution with **custom-built models** in Azure ML.	Azure Machine Learning and Power BI Service – Dataflows or Power BI Desktop	Access and consume all data models created or shared to apply on a Power BI dataset to enrich your data. It can also be shared with a business analyst to access and enrich datasets. **This will not be covered in this chapter.**	You have machine learning models in Azure that you would like to use in Power BI solutions.

Given the options for using AI functionality on the Power Platform, the following sections will help you understand the "why" and the "how" and give you some more details on how to get started with your own AI and Cognitive Services solutions leveraging Power Platform.

Power BI and Cognitive Services

Cognitive Services was first introduced to Power BI in early 2019. The Power BI suite provides a collection of APIs abstracting the machine learning models for Vision, Language, Speech, Decision, and Search capabilities. The services supported today are sentiment analysis, key phrase extraction, language detection, and image tagging.

Sentiment Analysis

One common use case involves a dataset containing written customer feedback and social media posts. For example, a municipality wants to examine a dataset to understand the attitude toward Pfizer vaccinations based on tweets during the COVID-19 pandemic. We would like to know if the attitudes of the writers in a given timeframe have a positive, neutral, or negative sentiment toward the brand, topic, product, or service – in this case, the topic of vaccination or the brand of Pfizer. The Text Analytics API sentiment analysis feature can help us evaluate a text and output a value (sentiment score) between 0 and 1 where 1 is more positive and 0 is more negative. Tweets can be written in many forms, lengths, and languages. To illustrate, we can inspect the following tweet:

> *This is ridiculous. I had a pretty rough 36 hours after my second Pfizer dose with nausea and severe muscle and joint pains, but I'd do it again in a heartbeat.* **Vaccination** *protects me and people around me. I'll gladly suffer what was, after all, not awful side effects for that.*

What would you guess the sentiment score of the tweet would be? All we need to have get started is the data. This can include the text. it might be a JSON file, a text file, or an Excel sheet. It can come from any of the 90 different supported connectors in Power Query Online including Files, Databases, Azure, Online Services, or other.

Just be aware if it's a local/on-premises data source, you need to set up a personal or enterprise gateway to connect and manage the refresh. In this example, we are leveraging a sample of 90,000 tweets (source: `www.kaggle.com/keplaxo/twitter-vaccination-dataset#vaccination2.csv`) extracted to a CSV file, fetched tweets from

March to June and uploaded to a blob storage. And of course it will make sense to have some additional data in addition to the text, for instance, topic, author, page, mentions/replies (is it conversational?), emojis, country, date, etc., to give the text sentiment score some more meaningful context and be able to compare before and after a given timeframe. The dataset looks something like the one in Figure 6-1, only with a lot more columns.

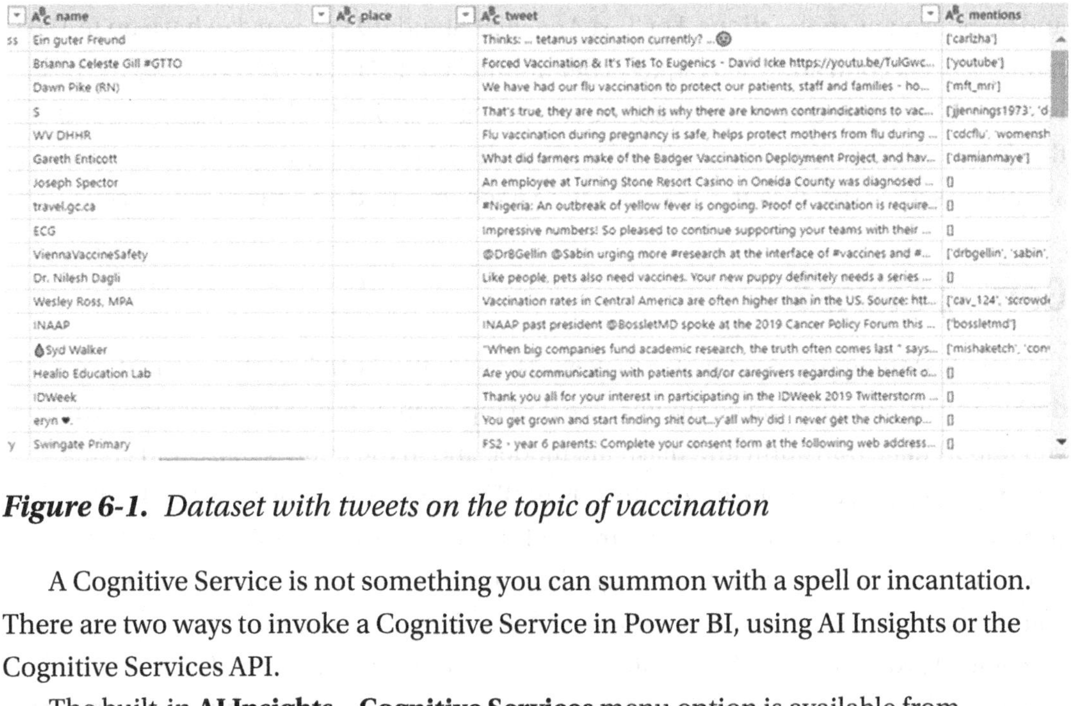

A⁂c name	A⁂c place	A⁂c tweet	A⁂c mentions
ss Ein guter Freund		Thinks: ... tetanus vaccination currently? ...😊	['carlzha']
Brianna Celeste Gill #GTTO		Forced Vaccination & It's Ties To Eugenics - David Icke https://youtu.be/TulGwc...	['youtube']
Dawn Pike (RN)		We have had our flu vaccination to protect our patients, staff and families - ho...	['mft_mri']
S		That's true, they are not, which is why there are known contraindications to vac...	['jjennings1973', 'd
WV DHHR		Flu vaccination during pregnancy is safe, helps protect mothers from flu during ...	['cdcflu', 'womensh
Gareth Enticott		What did farmers make of the Badger Vaccination Deployment Project, and hav...	['damianmaye']
Joseph Spector		An employee at Turning Stone Resort Casino in Oneida County was diagnosed ...	[]
travel.gc.ca		#Nigeria: An outbreak of yellow fever is ongoing. Proof of vaccination is require...	[]
ECG		Impressive numbers! So pleased to continue supporting your teams with their ...	[]
ViennaVaccineSafety		@Dr8Gellin @Sabin urging more #research at the interface of #vaccines and #...	['drbgellin', 'sabin',
Dr. Nilesh Dagli		Like people, pets also need vaccines. Your new puppy definitely needs a series ...	[]
Wesley Ross, MPA		Vaccination rates in Central America are often higher than in the US. Source: htt...	['cav_124', 'scrowde
INAAP		INAAP past president @BossletMD spoke at the 2019 Cancer Policy Forum this ...	['bossletmd']
🔵Syd Walker		"When big companies fund academic research, the truth often comes last " says...	['mishaketch', 'con
Healio Education Lab		Are you communicating with patients and/or caregivers regarding the benefit o...	[]
IDWeek		Thank you all for your interest in participating in the IDWeek 2019 Twitterstorm ...	[]
eryn ♥.		You get grown and start finding shit out...y'all why did I never get the chickenp...	[]
y Swingate Primary		FS2 - year 6 parents: Complete your consent form at the following web address...	[]

Figure 6-1. *Dataset with tweets on the topic of vaccination*

A Cognitive Service is not something you can summon with a spell or incantation. There are two ways to invoke a Cognitive Service in Power BI, using AI Insights or the Cognitive Services API.

The built-in **AI Insights – Cognitive Services** menu option is available from Dataflows, Power Query Online, or Power BI Desktop. This option requires content to be hosted on a Power BI Premium capacity that you have access to. Premium is one of the pricing models (in addition to Pro/Premium Per User licensing) Microsoft Power BI offers. The Premium SKUs offer resource tiers for memory and computing power. There are different options for Premium capacities, and maybe your organization already has invested in one. If you do not have a Premium capacity available and you would like to purchase one for testing purposes only, it's recommended to purchase a Premium Per User license or A2 Azure SKU for this scenario (A3 or A4 is dedicated, but on a fairly normal-sized dataset, A2 should be OK). The A SKUs can be turned off/paused when not in use, and thereby the Azure costs won't run crazy. Note that costs will run if you don't turn them off.

Initiate a custom call to the Cognitive Services APIs using Power Query, a built-in data import and transformation engine which uses the M language and is incorporated in in Power BI, Excel, and analysis services. Often times the M code is either generated for you using the UI or written by you. Using the cognitive service API requires an Azure subscription, as you will need to create a resource group and a Cognitive Services instance for Power BI to enrich your data. The cost of the Cognitive Services in Azure is based on usage, how often you need to call the Cognitive Services, how many records, and/or how many transactions. For Text Analytics Standard (S), you can have up to 100 requests per second and 1000 requests per minute. For 0–500,000 text records, the price is $2 per 1000 text records and $1 for the remaining 500,000–2,500,000 text records. If the dataset is larger, the next partitions will cost less. For testing, the free or standard tier should be sufficient. Of course the scale it up or down based on your specific needs.

Getting Started

The very first thing you need to do to get started with Power BI is to download the desktop client to your local computer. The desktop client is the "superuser" tool, where you import data, transform and mash up, model, and in the end visualize and create your reports – before publishing to the Power BI cloud service, where you will have some of the same capabilities, but far from all. There are two options. The first and best is to download the Power BI Desktop application from the Microsoft Store. That way you will get the updated product automatically each month. The second is to download Power BI Desktop from the Microsoft Download Center: `www.microsoft.com/en-us/download/details.aspx?id=58494`. For the second option, you can choose a language and either x32 bit or x64 bit depending on your operating system. Note that you will have to manually download when there are updates to the product using this option.

AI Insights in Power BI Desktop

In this example, the CSV file is imported into Power BI Desktop, using the **Get data** option and the **Text/CSV** data source and selecting the CSV file that stored on our computer. On the Power BI Desktop top ribbon, on the right-hand side, we see the out-of-the-box options illustrated in Figure 6-2 for AI Insights that we have available for us to select from.

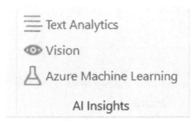

Figure 6-2. *Options for AI Insights in Power BI Desktop*

After we have selected the Text Analytics or Vision we will be prompted to log in to a dedicated Power BI Premium capacity to run cognitive services. The AI Insight options including Cognitive Services are Premium only, meaning that we will not be able to use them without a Premium capacity.

In this example, we have selected Text Analytics, logged in, and selected either the default one or a capacity of choice. We then see three options: Detect language, Extract key phrases, and Score sentiment. In this case, we will start by using the **Detect language** option as shown in Figure 6-3, so sentiment analysis can be run in the different languages supported.

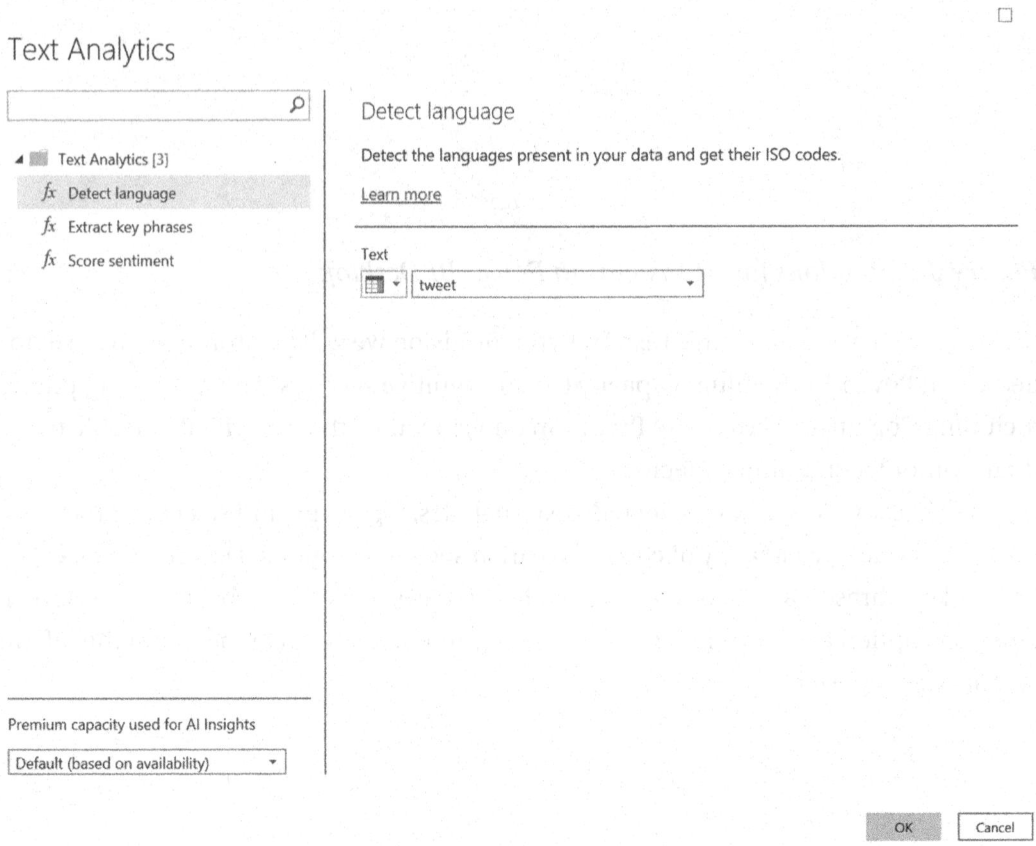

Text Analytics

Figure 6-3. *Detect language*

When we then select **Detect language** option, the next step is to select the column. We are going to select the column from the dropdown box, in this case "tweet," and click on the **OK button**. We immediately see that a function is created for us in the Queries pane on the right side in the **Applied Steps** pane as shown in Figure 6-4 which contains the step.

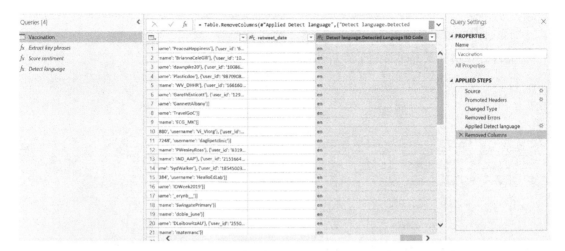

Figure 6-4. *"Applied Detect language" in Power BI Desktop's Power Query*

Once we have the ISO language code for the tweets in our query, there may be some language codes might not be supported in Cognitive Services APIs, like in this case, if we go directly to Close and Apply in the main ribbon, we will get an error on "<pii>tl</pii>" is not a supported language. However, that value was nowhere to be found in the dataset and could potentially be a bug. We can and should filter the dataset to include only those rows containing the supported languages, by clicking the dropdown on the column header, and text filter-advanced and add the languages and in or clauses like we see in Figure 6-5. If we skip this step, we will not be able to load the data in the following steps if an error occurs.

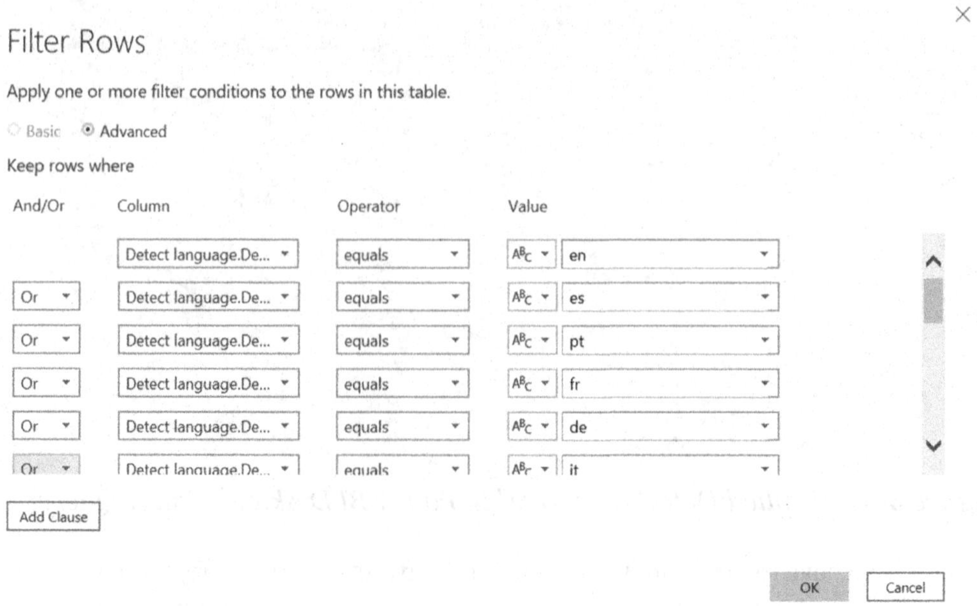

Figure 6-5. *Filtering out rows containing unsupported languages*

Once we have filtered out the unsupported languages, we will again select the AI Insights Text Analytics option in the ribbon just like we did with Detect language (Figure 6-3), but in this case we use the **Score sentiment** option. The text we would like to score will be the "tweet" column in the dataset, and we will include the ISO language code from the first step in Figure 6-5 and apply to the **Text** where the text Tweet is shown in Figure 6-6.

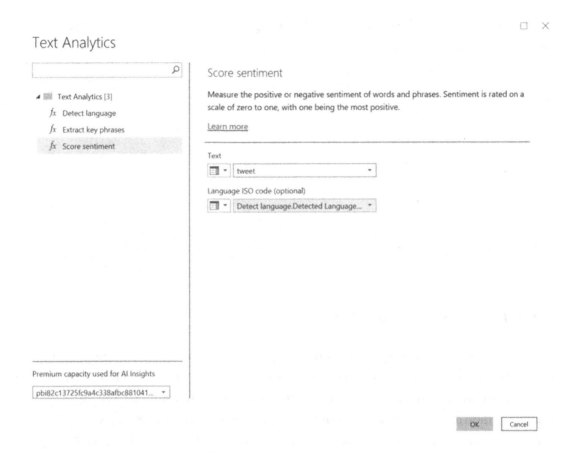

Figure 6-6. *Score sentiment function applied on the "tweet" column*

We will then see that a function is created for us as we have still not written a single line of M code which is used in Power Query, calling the Cognitive Services Score sentiment text API as you can see next to the blue arrow in Figure 6-7.

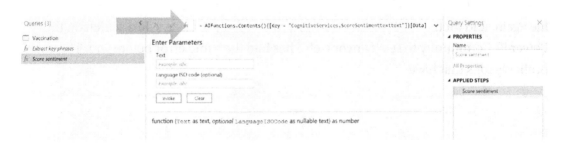

Figure 6-7. *Function calling the Cognitive Services Score sentiment text API*

And if we return to the Vaccination table in Figure 6-8, we can see that the Score sentiment column is now applied.

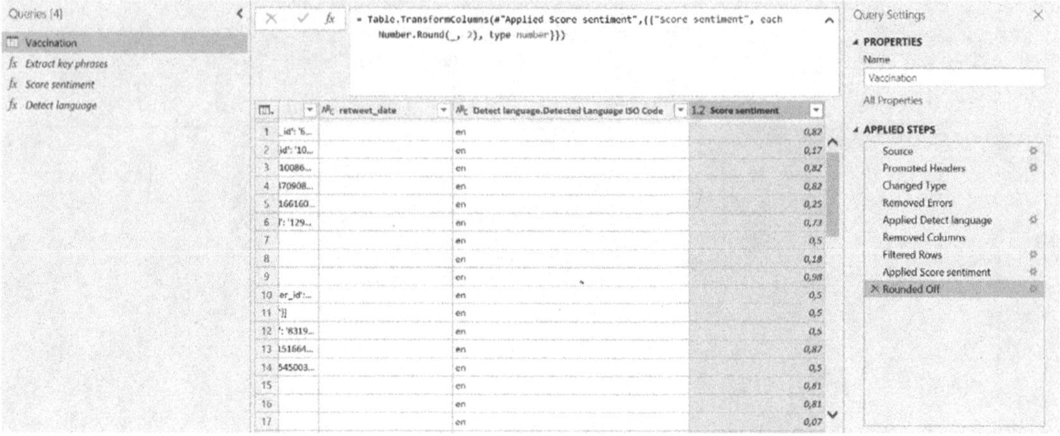

Figure 6-8. *Applied Sentiment score*

The function has been applied to the "tweet" column, and we have a new column containing the sentiment score of a tweet (Score sentiment), with a bunch of decimals initially. Yes, you are thinking correctly; it does makes sense to round off as a next transformation step (in transform rounding in the ribbon). We can also do other transformations to the sentiment output, for instance, add a new column with above 0.6 = "positive" and below 0.4 = "negative" for reporting reasons. Once we are happy with the dataset, we can apply the query and load the data into memory.

If errors happen during load – like the example in Figure 6-9 – check if your Premium capacity has reached its limit using Premium capacity metrics. The error might also be related to the new metadata format in Power BI Desktop introduced as preview in March 2020. The error will disappear if you turn of the new metadata format, open the file again, and then refresh the dataset. As a rule of thumb, always do a search on the Power BI Community to see if anyone else has had the same error before you. There are probably tons that have.

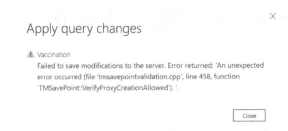

Figure 6-9. *Error that can occur on apply and load*

Start with AI Insights in Power Query Online

You can also create AI Insights and Cognitive Services from the Power BI Service in the Azure cloud. When you create the insights from Dataflows in Power Query Online, it's almost the same experience which is shown in Figure 6-10. The only difference is that there could be a few more options for Cognitive Services in the cloud service and a slightly different look and feel due to the release cadence of features being slightly more often and desktop + service not being in parity as of this date. However, one important distinction is that Dataflows queries are stored and run on Azure Data Lake – a scalable cloud data storage facility (either bring your own or use the default Microsoft lake). This enables datasets to be processed much faster than in Power BI Desktop. The Cognitive Services API also runs much faster in Dataflows. After refreshing Dataflows successfully, you can then connect to it in Power BI Desktop using Get data from Power BI Dataflows and create the same reports from Power BI Dataflows as a source.

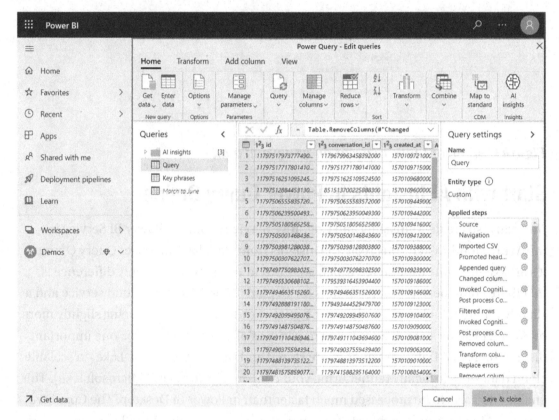

Figure 6-10. *Power BI Dataflows/Power Query Online experience*

Start with Azure Cognitive Services in Power BI

We can also leverage Cognitive Services APIs in Azure in Power BI Desktop. This will require an Azure subscription, but on the other hand, no Premium capacity is needed. The APIs we then can use are more than the services built in using "AI Insights."

The first thing we need to do is to set up the Cognitive Service in Azure Portal. Searching for Cognitive Services gives us plenty of options to choose from. We will use the same dataset as described in the preceding and therefore choose to create a Text Analytics Cognitive Service as shown in Figure 6-11.

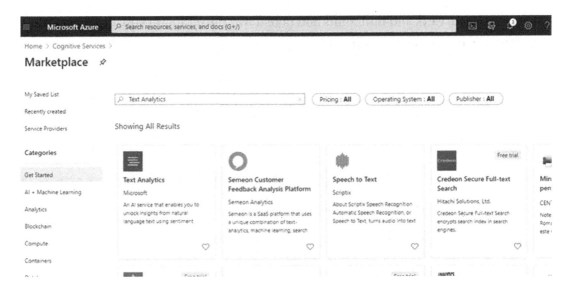

Figure 6-11. *Marketplace in Azure – Cognitive Services for Text Analytics*

We place the Cognitive Service in a location nearest us, and it's also wise to choose the same region where Power BI is placed to avoid latency. To find out what location the Power BI Service is located, from within the Power BI Service, click on the question mark icon on the top right side of the screen and select **About Power BI** to find the location. We can choose the lowest pricing tier for now. We then go to Quick start, go to API Console (V2) in the second option as shown in Figure 6-12, and note the location URL for the Cognitive Service.

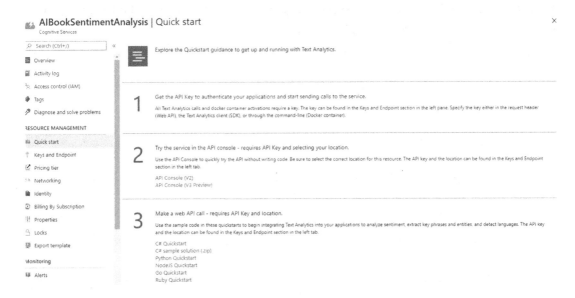

Figure 6-12. *API console to get the location URL for the Cognitive Service*

We then go to Keys and Endpoints and note down the endpoint URL and Key 1 as shown in the screenshot in Figure 6-13.

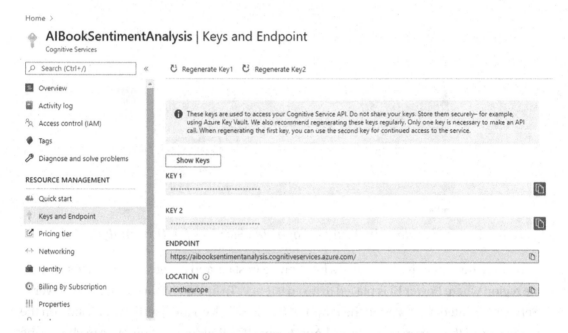

Figure 6-13. *Keys and Endpoint for the Cognitive Service in Azure*

We are then ready to go into Power BI Desktop to call the API in Power Query on the same dataset used in the previous sections. First, we need to create a blank query (new source, blank query) as shown in Figure 6-14.

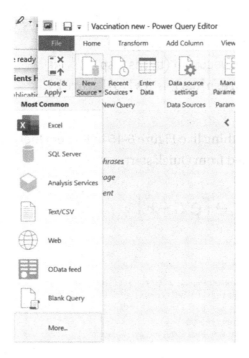

Figure 6-14. *New Power Query*

From there, it's easiest to open the advanced editor and paste in the M function shown in Listing 6-1, only replacing the endpoint and the API key with the ones we got from Azure.

Listing 6-1. M Code to Be Pasted into Blank Query in Power BI Desktop

```
(text) => let
    apikey     = "<Your API Key 1>",
    endpoint   = "https://< Your Text Analytics API (V2.1)>/text/
                 analytics/v2.1/sentiment",
    jsontext   = Text.FromBinary(Json.FromValue(Text.Start(Text.
                 Trim(text), 5000))),
    jsonbody   = "{ documents: [ { language: ""en"", id: ""0"", text: " &
                 jsontext & " } ] }",
    bytesbody  = Text.ToBinary(jsonbody),
    headers    = [#"Ocp-Apim-Subscription-Key" = apikey],
```

```
    bytesresp   = Web.Contents(endpoint, [Headers=headers,
                    Content=bytesbody]),
    jsonresp    = Json.Document(bytesresp),
    sentiment   = jsonresp[documents]{0}[score]
in  sentiment
```

This would look something like Figure 6-15 in Power Query, and note that the endpoint is the one fetched from Quick start.

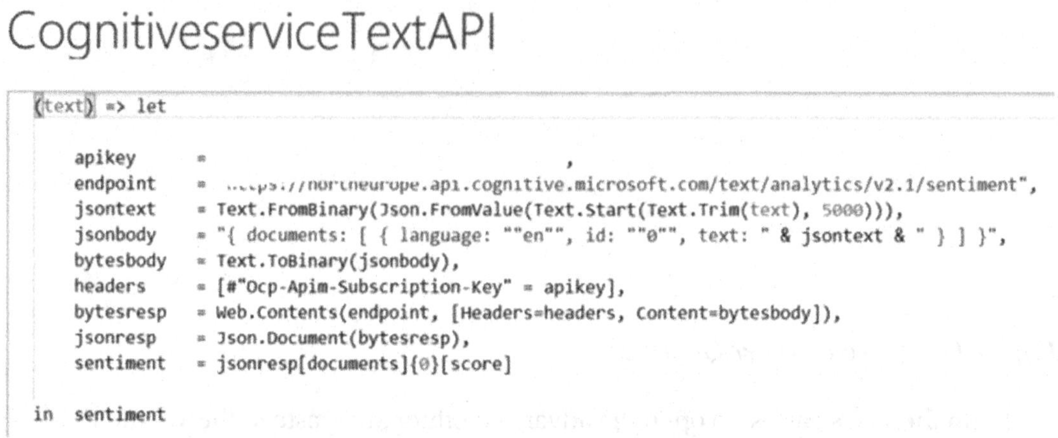

CognitiveserviceTextAPI

```
(text) => let

    apikey      =                              ,
    endpoint    = ...ps://norcneurope.api.cognitive.microsoft.com/text/analytics/v2.1/sentiment",
    jsontext    = Text.FromBinary(Json.FromValue(Text.Start(Text.Trim(text), 5000))),
    jsonbody    = "{ documents: [ { language: ""en"", id: ""0"", text: " & jsontext & " } ] }",
    bytesbody   = Text.ToBinary(jsonbody),
    headers     = [#"Ocp-Apim-Subscription-Key" = apikey],
    bytesresp   = Web.Contents(endpoint, [Headers=headers, Content=bytesbody]),
    jsonresp    = Json.Document(bytesresp),
    sentiment   = jsonresp[documents]{0}[score]

in  sentiment
```

Figure 6-15. *Function in Power Query calling the Cognitive Services API in Azure*

The language can also be changed to not be hard-coded to "en" and instead replaced with the ISO code we already have in our dataset. That is a best practice: never use sentiment analysis "en" on texts that are in other languages.

Then it's time to invoke the custom function option on the vaccination dataset tweet column as shown in Figure 6-16.

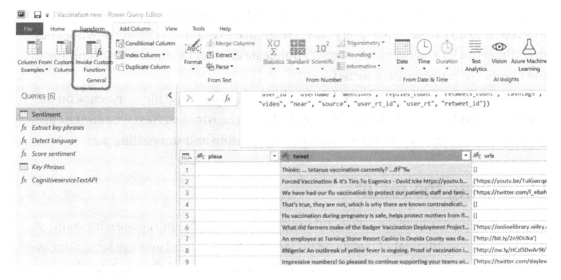

Figure 6-16. *Invoke Custom Function button in Power BI Desktop*

This will create the new column for the invoked function as shown in Figure 6-17.

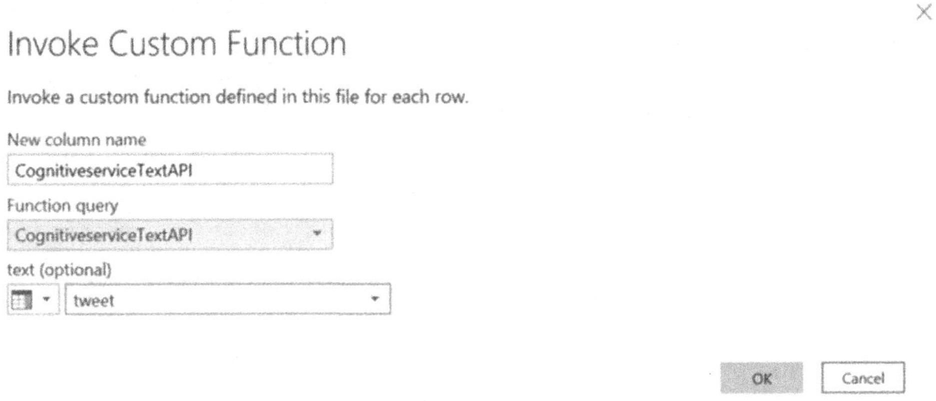

Invoke Custom Function

Invoke a custom function defined in this file for each row.

New column name

CognitiveserviceTextAPI

Function query

CognitiveserviceTextAPI

text (optional)

tweet

OK   Cancel

Figure 6-17. *Invoke Custom Function user interface in Power BI Desktop*

When we click OK, we will load the data into memory. It will by default call the API row by row, which can be quite slow if you have a large dataset. It's therefore recommended to do it in batch mode, which is explained by Gil Raviv in this blog: `https://datachant.com/2016/08/09/sentiment-analysis-power-bi-part-2/`. I'd also like to thank Marc Lelijveld that has written a blogpost on the topic earlier this year. There is also very good documentation on invoking Cognitive Services APIs on the Microsoft Docs pages if you search on Power BI and Cognitive Services.

Before we move on, we can also extract the key phrases from the tweet; however, this is best to do in a separate query (use a reference to our main dataset and filter out everything but ID and tweet) since it will duplicate all rows in our dataset according to the number of key phrases detected.

To apply the function, click **OK** button. To load the data into memory, click on the **Close and Apply** button in the ribbon on the left of the window. Next we can start the interpreting the results and move on to the visualization and storytelling part.

Visualizing the result

To visualize the results in a meaningful way, we can create a report page with count of tweets, average sentiment of tweets related to #vaccination, and average sentiment over time, along with the actual tweets. We can drill down, filter, and slice and dice based on period and sentiment as shown in the example in Figure 6-18.

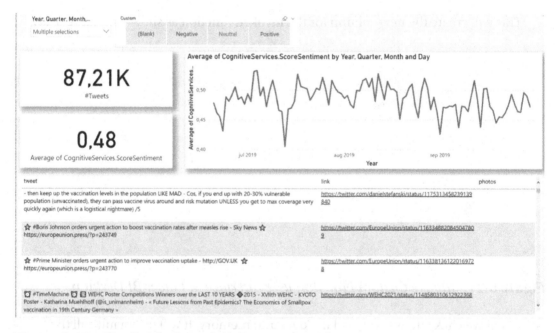

Figure 6-18. *The finished result in Power BI Desktop*

The next thing we need to do is to deploy the report to a Power BI Service workspace and create an app to distribute content to the relevant end users and set a refresh schedule on the dataset. For the preceding case, it's recommended to first apply incremental refresh to only update the last incoming rows (ten days for instance) instead of refreshing the entire dataset every time.

To learn more about the algorithm used in this example, see `https://blogs.technet.microsoft.com/machinelearning/2015/04/08/introducing-text-analytics-in-the-azure-ml-marketplace/`.

AI Visuals

AI visuals in Power BI are also include Power BI as part of the standard visuals and were added in 2019, so they haven't been around for that long. This is another example of Microsoft is investing heavily in AI being built across their ecosystem and further embarking on the vision in enabling everyone to become data driven and even perform complex tasks and interpret or analyze data in easy (easier) ways than ever before. Another fun fact for you: AI features are even backed into some of the traditional visualizations in Power BI, such as bar and column charts. Now what kind of AI visuals do we have to choose from, and how can we use them?

Explore Data with the Q&A Visual

The Q&A visual is great for data exploration. It enables you to ask questions related to your data in more or less a natural language. The Q&A visual can be set up by reviewing questions people have asked, for instance, "How many vaccination tweets were there with negative sentiment in March?" The visual also allows a report developer to predefine questions for the end users and to fix any misunderstandings in terms of questions asked and data returned in the visual.

The example shown in Figure 6-19 goes back to the vaccination sentiment topic introduced earlier. We can use the Q&A visual to discover the dataset and auto-visualize data by asking questions like

- What is the average sentiment?

- What is the number of Tweets By Sentiment?

- What is the number of Tweets By Sentiment over time?

Questions can be suggested, to get the users started, but in general they can come up with their own questions in the "Ask a question about your data" input field.

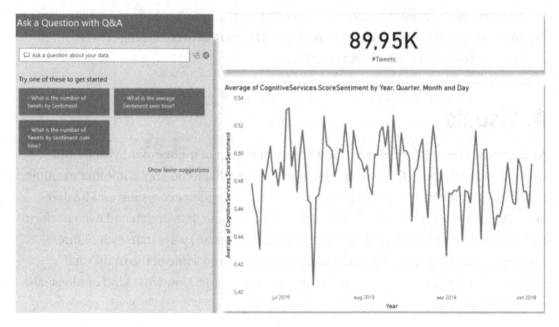

Figure 6-19. *Ask a Question with Q&A menu on a report page*

The suggested questions you see in Figures 6-19 and 6-20 can be configured by a report developer like yourself.

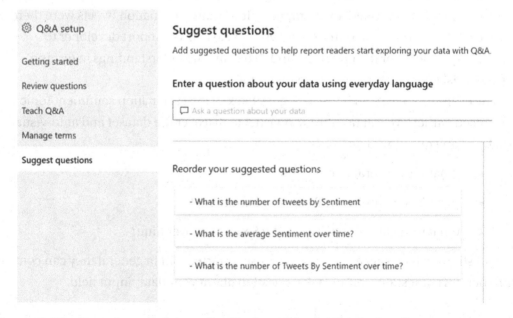

Figure 6-20. *Suggest questions to be displayed in the Q&A visual*

When choosing or asking a question, the user will get back an autogenerated visualization based on what Power BI thinks is best to visualize the data, like the one in Figure 6-21: "What is the number of Tweets by Sentiment over time?"

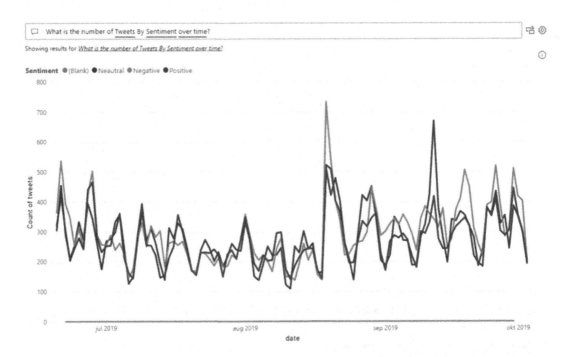

Figure 6-21. *Q&A visualization*

In this case, we get back a line diagram with a legend for Neutral, Negative, or Positive tweets from July to October 2019. If another question was asked, like average sentiment of tweets, we would get back a card visual showing one number.

Q&A can also be "trained" to understand questions and to understand and adapt to company char gong, synonyms, and domains. As shown in Figure 6-22, Q&A will identify terms that are not understandable, like "question" as shown in this example. We can then map "question" to describe our data. For instance, I would like "account" to be the same as "usernames" in our dataset. The mapped terms can be changed and managed later in the Q&A setup.

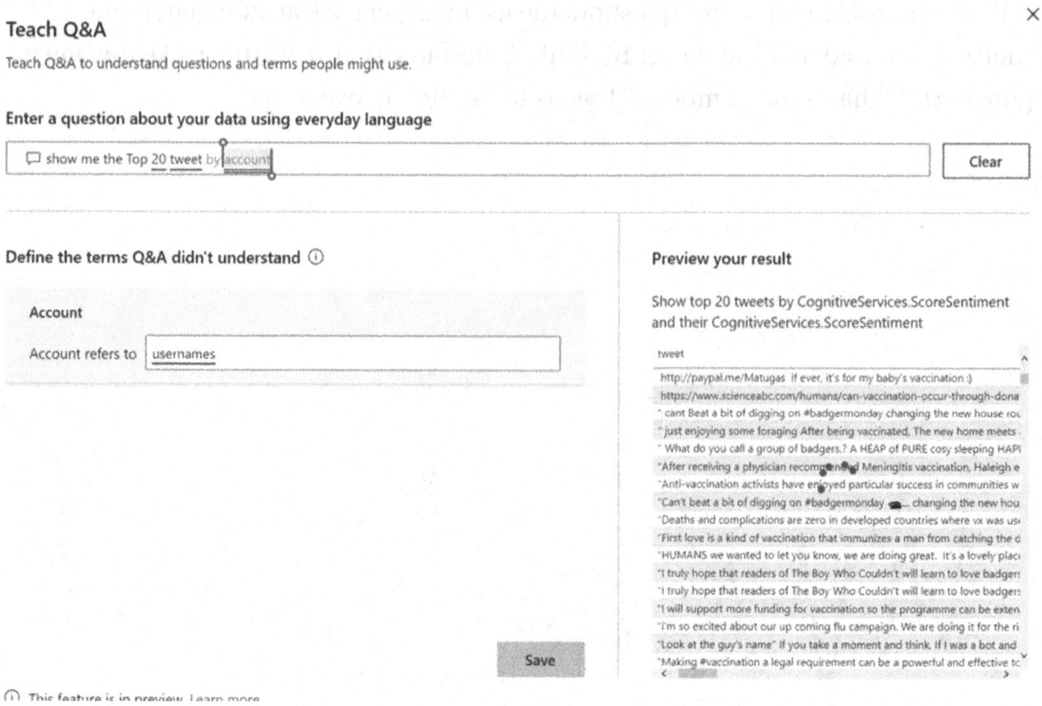

Figure 6-22. *Teach Q&A to understand questions in a more natural way*

Identify Key Influencers

Another popular AI visual is the key influencers visual. The visual lets us look at patterns in the data for one group compared to another group. One way you might use this visual is to look for Twitter accounts or countries that have a particular negative attitude against vaccination. The Key Influencer visual can answer questions such as What influences users to tweet content with low sentiment? What influences Airbnb users to rate a house rental low? The visual then lets us find patterns across the dataset, for instance, themes, subscription types, geography, count of replies, etc.

Going back to the dataset we have used throughout the chapter, on tweets, we might want to know what is potentially influencing sentiment to be negative. Then the key influencers visual is perfect for us to explore the data further.

First, we need to configure what we would like to be analyzed and explained by which dimensions. We choose the key influencers visual from the Visualization pane in Power BI Desktop and configure what we want to analyze by which explanation factors as shown in Figure 6-23.

Figure 6-23. *Field pane for configuring the visual*

In this case, we have chosen to analyze sentiment by date, name, which users the tweet is replying to, which users are mentioned in the tweet, username of account, user ID of account, and time. There was only a hit on the date dimension in this case, and we can see in Figure 6-24 that the likelihood of sentiment being negative increases by 1.09 if the Month is September and 1.08 if the Month is June.

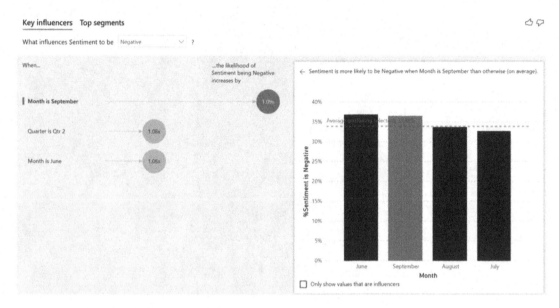

Figure 6-24. *Key influencers for negative tweets*

The orange average line in Figure 6-24 is showing us how it compares to all other months, which also indicates that the difference is not that big from the other months. But the data indicates that we maybe should look deeper into the months of September and June. It's also a good idea to play around with what type of data in our dataset could explain something, by testing and dragging in new columns in the "Explain by" field pane.

Analyze Root Cause with the Decomposition Tree

If we continue exploring our Twitter vaccination dataset, it might be valuable to know the key Twitter accounts or hashtags that are associated with "negative" sentiment. Perfect, we can use the decomposition tree to do some real root cause analysis on our dataset, enabling us to drill down into dimensions to help us explain a metric in any order. Slicing data starting from number of tweets and further slicing and drilling into the data into a tree structure by choosing a category like sentiment, we can break the data further down into the Twitter accounts that have negative sentiment and all the way to the hashtags used and in the end the tweets. The configuration of this breakdown is shown in Figure 6-25. This enables users to slice and dice data from the dataset as they see fit and perform AI Insights for high or low values. The AI Insight part comes into play whenever you see a light bulb.

Figure 6-25. *Configuration of the decomposition tree visual*

What we need to do first is to configure the visualization, with what we want to analyze and how we want to explain it by categories and dimensions in our dataset. In this example, we want to visualize and analyze and slice and dice the count of sentiment, which is basically the number of tweets we have in our dataset. This must be a measure or an aggregate. Then explain by negative, neutral, or positive sentiment, name of Twitter account, hashtag, and tweet. The decomposition tree result is displayed in Figure 6-26.

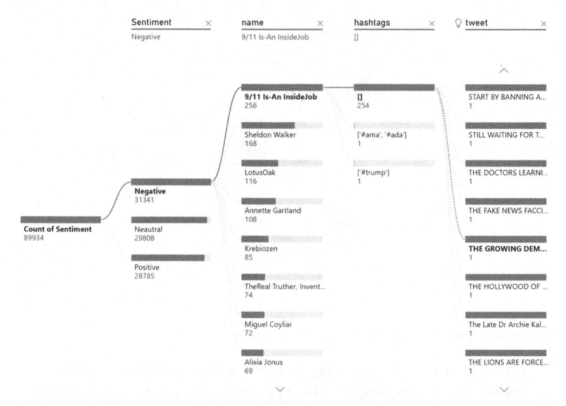

Figure 6-26. *Decomposition tree for the vaccination dataset*

Looking at the results in Figure 6-26, we see how many negative tweets "9/11 Is-An InsideJob" has and which hashtags they are using – in this case, they are not using hashtags that often, but we see two hashtags #ama, #ada, and #trump. Drilling further into the tweets with no hashtags, we see there is a light bulb next to tweet (see Figure 6-26 – last drill-down on the tweet), meaning that an AI split has kicked in recommending and highlighting in the visual which tweet has the highest value. This is then typically a value that we would be interested in looking into.

This can potentially give us some powerful insights, by breaking down our dataset into chunks, explaining increase or decrease across a set of categories we are interested in. Powerful, right?

AI Builder

AI Builder was launched in preview in June 2019 and is an even younger member of the AI in Power Platform family than AI in Power BI. AI Builder's purpose is to add intelligence to your organization's PowerApps and Power Automate with a point-and-click experience. The type of intelligence you can add is for the most common scenarios as follows:

- Form processing to pull data out from forms. Example: Scan invoice and pull out recognizable information such as date or company.

- Object detection to determine an object from a photo. Example: On the shelf inventory in retail/merchandise or scanning a business card.

- Prediction to perform binary analysis on data. Example: Predict value of account and special deals for marketing.

- Text classification to analyze data from text fields. Example: Sentiment analysis and routing customer feedback to the right team.

And by being a low-code platform, the application development costs and effort are said to be 70% less (total economic impact report...).

Start with a Prebuilt AI Model

The prebuilt AI models are the easiest to use, so let's start with that first. Let's take a similar example to the preceding one related to the COVID-19 crisis. Many organizations have a need to understand the well-being of their employees and maybe send out surveys to track how well they coped with the crisis when everything was shutting down and co-workers could no longer be working in a common office space and some even had children at home while working from home. Let's create a simple feedback text field where users can enter their reply to "How are you feeling today?" and then use PowerApps and a prebuilt sentiment analysis model to provide back if the sentiment of the free text comments is negative, positive, or neutral.

What we need includes

- A SharePoint list to store the feedback

- A PowerApp where the feedback can be input. If starting from
  SharePoint as a data source, then you will get the app ready made for
  you. You only need to add a label for the sentiment of the text input
  by the user

- AI trial for 30 days or AI license as well as a PowerApps license or
  E3–E5 Office 365 license

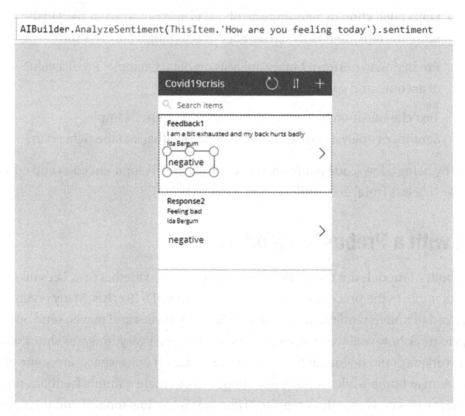

Figure 6-27. *AnalyzeSentiment function for AI Builder in a PowerApp*

The PowerApp in this example is created with SharePoint as a data source and will
look like Figure 6-27. The only line of "code" to get the sentiment output in the gallery is
the one displayed at the top of the function window in Figure 6-28, where we are taking
the item text displayed and using AI Builder to give back the sentiment.

In addition, we can add other functions to our gallery, like getting a list of standard categories detected in the feedback. This only requires the "CategorizeText" function as you see in the formula pane in Figure 6-28.

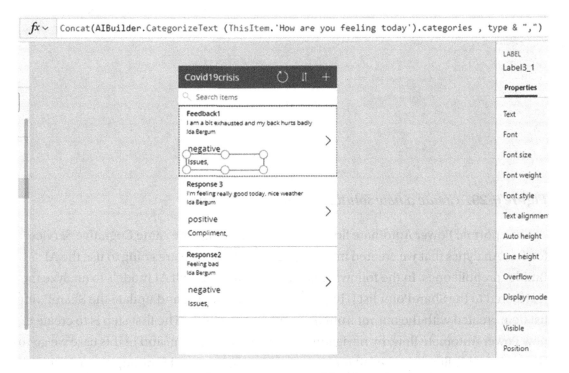

fx ∨   `Concat(AIBuilder.CategorizeText (ThisItem.'How are you feeling today').categories , type & ",")`

Figure 6-28.   *CategorizeText function in AI Builder for a PowerApp*

Let's take a small example on how AI Builder prebuilt models can be used in Power Automate as well. So let's start creating a new solution shown in Figure 6-29 and add a flow to our solution.

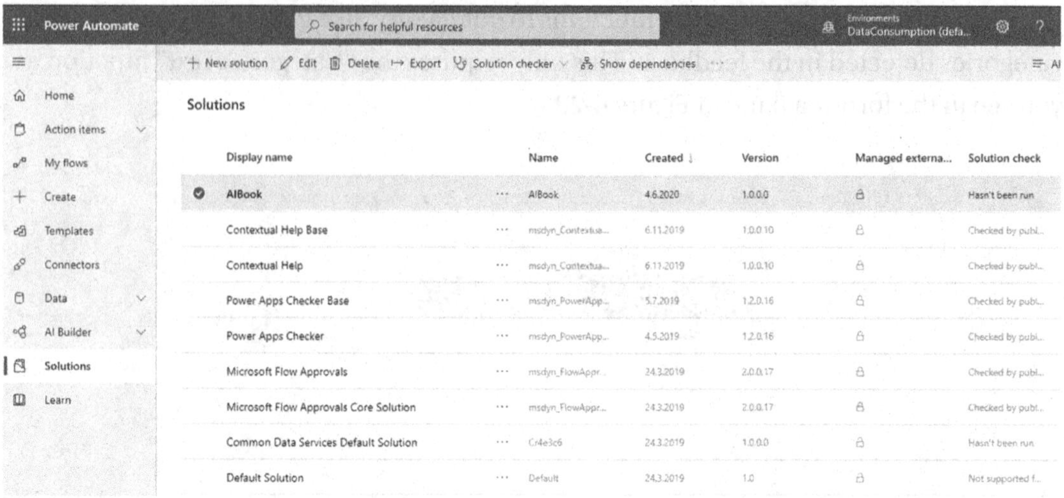

Figure 6-29. *Create a new solution in Power Automate*

In a normal Power Automate flow, we can also leverage the same Cognitive Service for Text Analytics that we created in Azure, but in this case we are going to use the AI Builder prebuilt ones. In the following flow, we use the prebuilt AI models to analyze the text stored to the SharePoint list (How are you feeling today?) and update the SharePoint list item created with the output from the sentiment analysis. The first step is to create a new Power Automate flow, by navigating to flow.microsoft.com, and in this case we go to "Create" or "My flows" in the left menu, and a blank automated flow pops up, where we choose "When an item is created SharePoint" to be the trigger as shown in Figure 6-30.

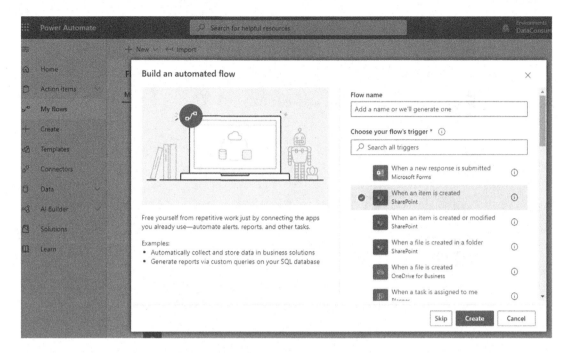

Figure 6-30. *Create a new Power Automate flow for when an item is created in SharePoint*

We then configure the flow created for us, to use the Analyze positive or negative sentiment in text AI Builder function. And store the sentiment to the SharePoint list for each new entry like what we see in Figure 6-31. That way we also have the data stored and not only displayed like we did in the PowerApp.

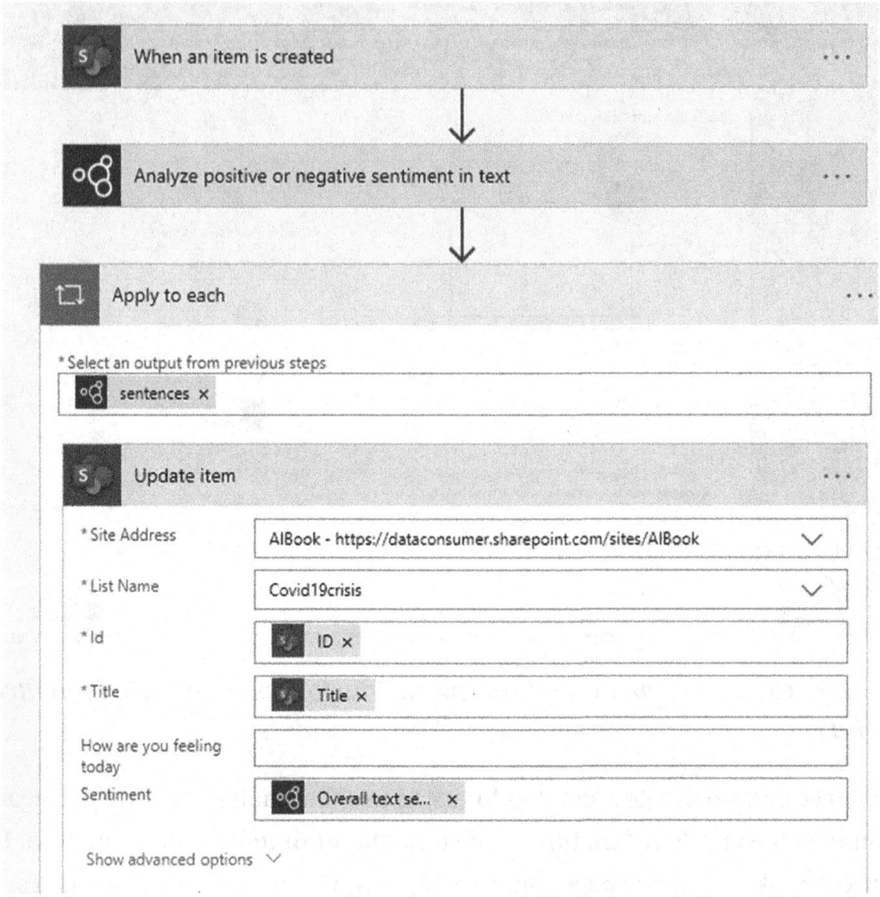

Figure 6-31. *Power Automate flow using AI Builder for items created in a SharePoint list*

Running the flow in Figure 6-31 will then get us the sentiment of the feedback every time it's provided by users, stored in our SharePoint list in Figure 6-32 (or any other storage supported for that matter), so we can reuse the data for reporting in Power BI or integrate the data to other apps or services.

Figure 6-32. *A response in the SharePoint list with sentiment from the Power Automate flow*

There are many more models to choose from. Some are only available in Power Automate, and some are available in both PowerApps and Power Automate. Two examples include the business card reader and text recognition in the Vision category. There are more language-specific models in addition to category classification and sentiment analysis like entity extraction, language detection, and key phrase extraction (these we have used in the Power BI examples). To get a good overview of models, you can have a look here for the complete documentation https://docs.microsoft.com/en-us/ai-builder/model-types.

Customize Your AI Model

In addition to the prebuilt models, we have the option to customize models. This is where you choose which AI model suits your business need from the models shown in Figure 6-33.

Figure 6-33. *Customize AI models for PowerApps and Power Automate*

The models can be applicable for the various scenarios summarized in Table 6-2.

Table 6-2. *An Explanation of the Models*

Model	Use Case Example
Category classification	One of the fundamental natural language processing challenges is identifying textual data with tags. These tags can be used for sentiment analysis, spam detection, and routing customers from textual data coming from email, documents, and social media.
Entity extraction	Identify key elements in text and classify into predefined categories. Helps you transform unstructured data into machine-readable structured data. You can then process the data to retrieve information, extract facts, and even answer questions.
Form processing	Identify and extract key/value pairs and table data from form documents. Some of the things you can use it for are as follows: Autodetection and validation of fields detected from invoice is a very good use case. Business card reader is another good one, more in the Personal category.
Object detection	A model that lets you upload images and tag them, so that once a scan occurs for that object with your phone using a PowerApp, the tags you have set will be displayed. Good use cases are Detecting parts and providing back repair manual in manufacturing Detecting a product in a shelf and providing back inventory in retail
Prediction	Analyze yes/no patterns in your data and learn to associate historical patterns in your data with outcomes. The model learns patterns based on past results and detects new data coming in to predict future outcomes. Questions you might have are Which customers are more likely to churn? Which leads are more likely to purchase a product or service?

After selecting which model suits your use case, you connect to your data from the Common Data Service. You tailor your AI model by filtering data, schedule it, and tweak the AI model to meet your needs and optimize how the model performs. Then it's time to train the model, which is an automatic process based on your tweaks. And in the end, the goal is to use the results of the AI model across Power Platform to create even more powerful solutions for your business needs. Power Automate can, for instance, be used to automate document processing.

Let's review an example for a prediction model – Will the customer make a purchase? –using binary prediction and numerical prediction based on web analytics data from an online retailer.

To get started, we need to download a sample dataset solution called "onlineshopperIntention" from the following link: `https://go.microsoft.com/ fwlink/?linkid=2093415`. We need to import it into a custom entity in the formerly known Common Data Service (CDS), now rebranded to Datavers. It is commonly known as CDS, and is a service on Azure Data Lake, almost like writing to and from a database. It lets us store data securely in the cloud in so-called "entities." CDS has a set of commonly used entities like "Account" and "Customer" with related metadata which also includes an option to create "custom entities," when they are needed.

Before importing the data, go to make.powerapps.com and navigate to the "Solutions" option on the left-hand side at the bottom. When you are on the Solutions screen, there is the option to choose **Import** that we would like to use. When we click **Import,** a new window will open, enabling us to select a solution package which is shown in Figure 6-34. We will then choose the .zip file we just downloaded in the first step.

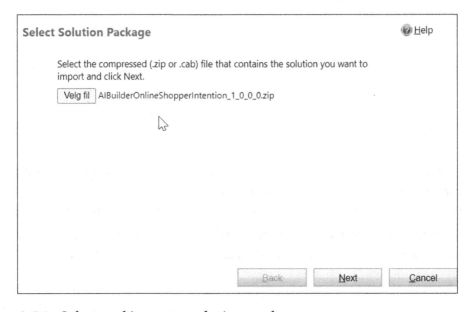

Figure 6-34. *Select and import a solution package*

When importing, we will get an info warning that we are importing an unmanaged solution. We will ignore that for now, since we do not have any existing solution at risk of being overwritten, and choose to import. It should go smoothly and say "Completed successfully," and we will get a preview of the import. We can then close the window, and we should see "AI Builder Online Shopper Intention" in our solutions. And if we click it, we will navigate to the entity where we next will import the sample data "aib_onlineshopperintention.csv" file from the same link we downloaded the solution .zip file. We will then see the window as shown in Figure 6-35.

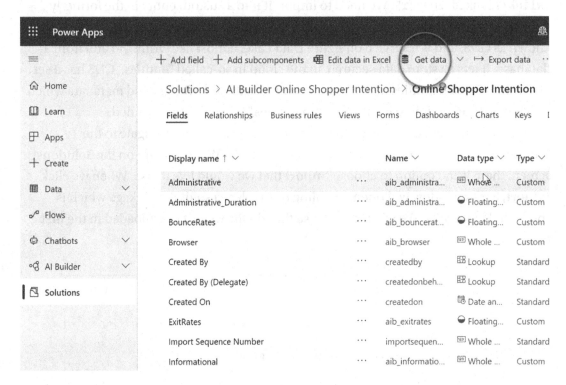

Figure 6-35. *Get data*

We would like to click "Get data" that will open Power Query Online (this might be familiar from when you worked in Power BI earlier – you are correct, it's the same UI). Select Txt/CSV; that is where we will get our data from. Copy the URL `https://raw.githubusercontent.com/microsoft/PowerApps-Samples/master/ai-builder/aib_onlineshopperintention.csv` into the file path or URL field like in Figure 6-36.

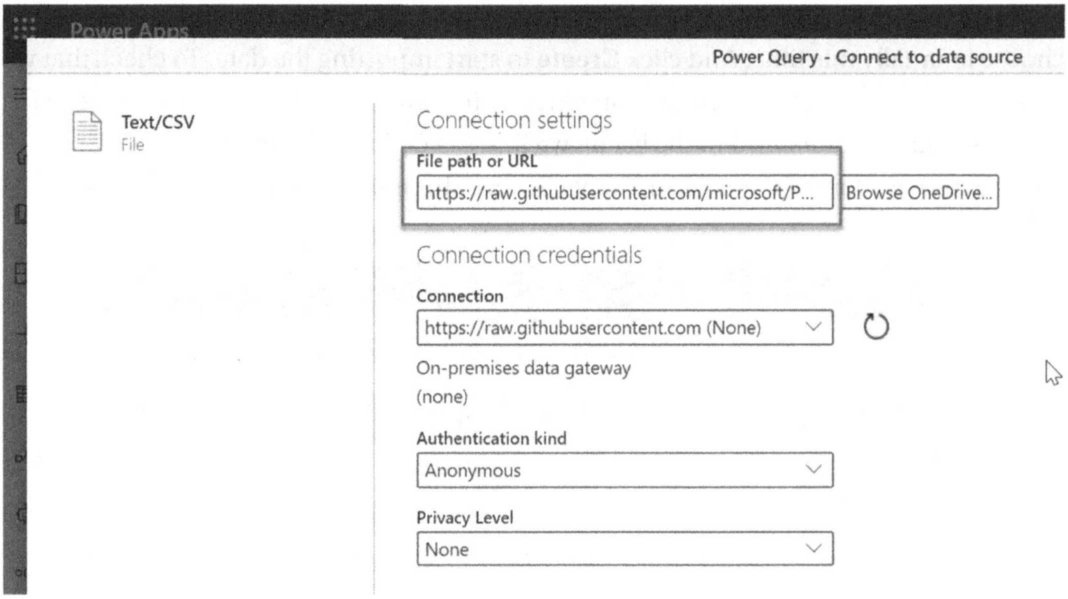

Figure 6-36. *Import a CSV file*

When we click "Next," we'll get a preview of the data based on the first 200 rows. We can click "Next," which will take us to the transformation step. We don't need to transform any data as of now – only check that it's comma delimited. So we can click "Next" again – and we will end up in the Map entities window as shown in Figure 6-37. We will select "Load to existing entity" and select the aib_onlineshopperintention entity in the dropdown list. We would also like to use the "Auto map" button, so Power Query will automatically map source columns to destination fields based on the name.

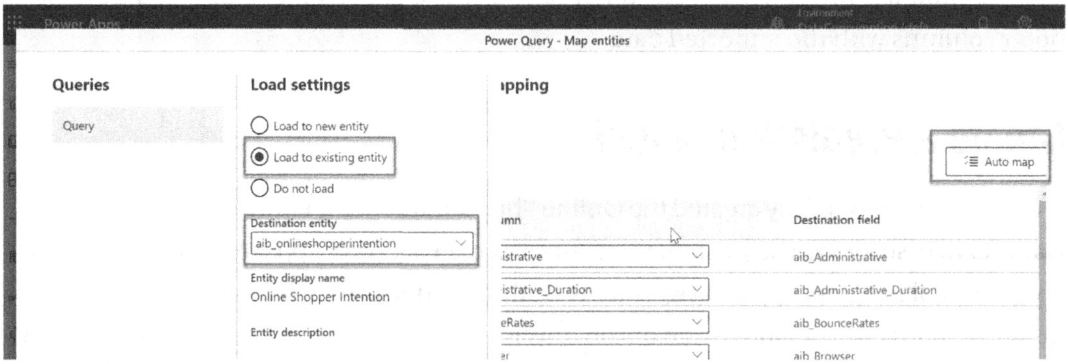

Figure 6-37. *Auto-map entities*

Next, we will get the option to refresh settings, manually or automatically. We will choose manually this time. And click **Create** to start importing the data. To check that we have imported what we wanted, we can navigate to **Data** on the left-hand menu, select **Entities**, and navigate down in the list till we find the Online Shopper Intention. Go to Views and select one of them – like "Active Online Shopper Intentions" in Figure 6-38.

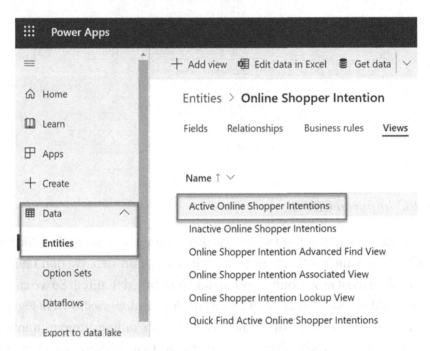

Figure 6-38. *Prediction model entity views in AI Builder*

You should now see a list of customer names, the Created on column, and a bunch of other columns with the imported data.

Create a Prediction Model

We have now successfully created the Online Shopper Intention entity in the Common Data Service, so let's use it for something fun! To create a PowerApps prediction model. If you are still in PowerApps, navigate to **AI Builder** and **Build** and select to refine the model for prediction. Before we go ahead, this is where you really need to consider which prediction you want AI Builder to make based on your machine learning problem or questions you would like to have answered. Are you interested in knowing if a customer is about to churn or not? And then maybe you can offer an incentive to the ones about to churn. Are you interested in knowing why they are leaving – and maybe there is a field that states if the customer has churned? You also need to get an understanding if there are any fields in your dataset that can cause any uncertainty in the results.

Once we have those questions sorted out, we know in this case that we have many customers with no revenue. And we would like to know if a customer that has interacted with the online store did make a purchase and predict the likelihood of customers making a purchase in the future. Let's select the entity "Online Shopper Intention" and the historical field "Revenue (Label")" in Figure 6-39 to predict the outcome.

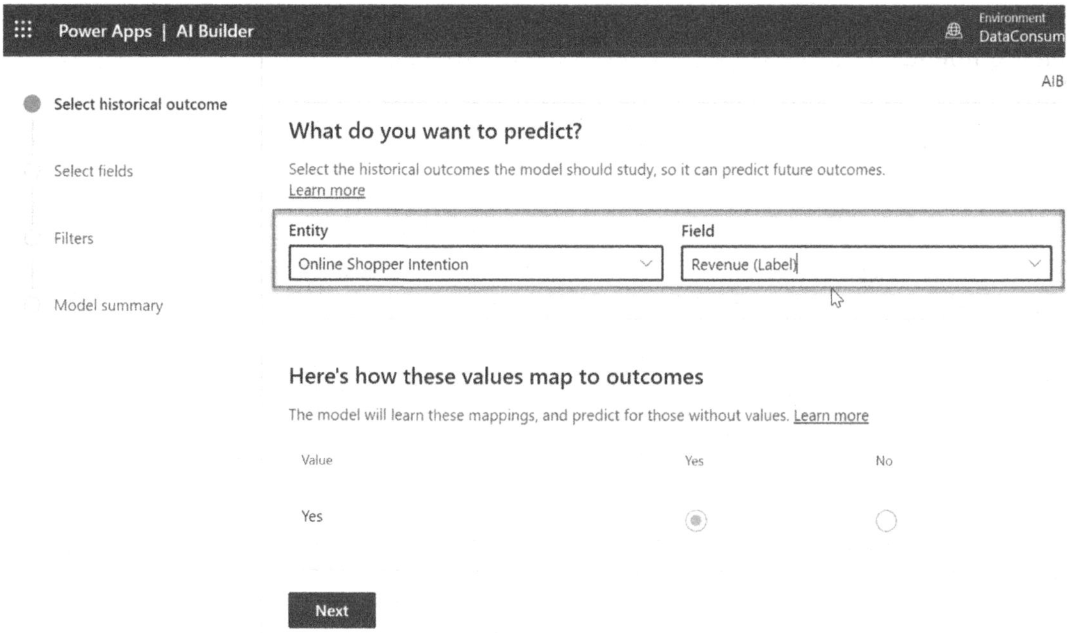

Figure 6-39. Prediction model outcome

When we click Next, we will train the data. By default, all the relevant fields are selected, but we can choose to deselect the ones we don't want, and that might end up contributing to a less accurate model. Take some time when working on your own data to think through your scenario. In our case, we will let AI Builder select for us. Next, we will have the option to filter our model, like looking at customers from a specific country or region only (like where the outcome is unknown for only region US). We can filter on a row, a group, or a related entity. We'll skip that for now, and next we will select to train the model just like we do in Figure 6-40. This can take a little bit of time, so go have yourself a well-deserved coffee break!

Models

My models   Shared with me

°⅋	Name	Last trained ↓	Permission	Owner	Status
⌁	AIBookPredictCustomerBuy	Training	Owner	Ida Bergum	○ Training

Figure 6-40. *Train the model*

After the model has been trained, we are happy with the model after interpreting the results, and tuning in Figure 6-41, we can publish it to make it available for users in your environment.

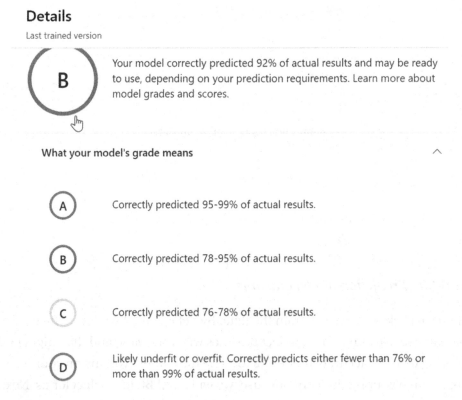

Details

Last trained version

B   Your model correctly predicted 92% of actual results and may be ready to use, depending on your prediction requirements. Learn more about model grades and scores.

What your model's grade means   ∧

A   Correctly predicted 95-99% of actual results.

B   Correctly predicted 78-95% of actual results.

C   Correctly predicted 76-78% of actual results.

D   Likely underfit or overfit. Correctly predicts either fewer than 76% or more than 99% of actual results.

Figure 6-41. *Interpreting results*

Use the Model

We can now use the model across Power Platform, in a canvas app, or in the formula bar just like we did with the analyze sentiment example. Or we can create a model-driven app and a view of our data with predicted outcomes, using the PowerApps Designer and dragging in the Online Shopper Intention in the view section in Figure 6-42 and selecting the fields of interest, saving, and publishing.

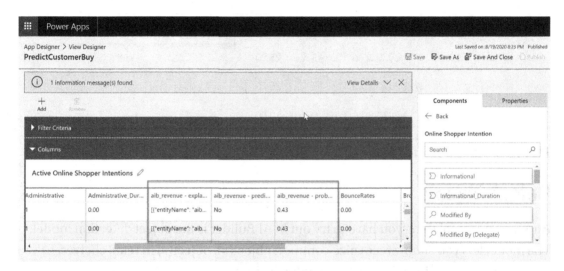

Figure 6-42. *View with predicted outcomes – Designer*

When publishing the view, we can see that we have a 43% probability for customer 1 to not buy a product when visiting the online store. We can also have a look at the explanation field in Figure 6-43, to see which attributes are weighted in the model.

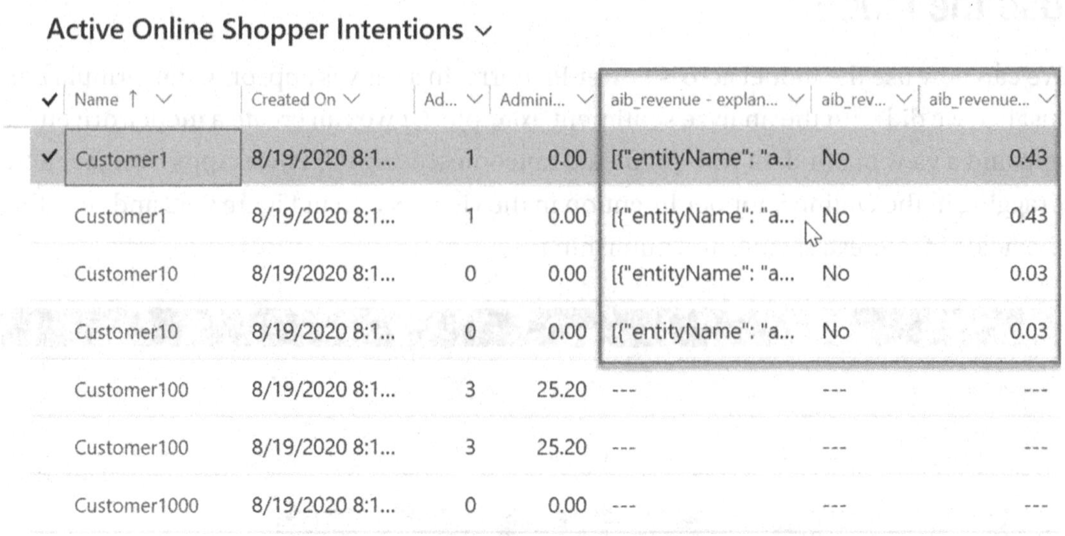

Active Online Shopper Intentions ∨

	Name ↑ ∨	Created On ∨	Ad... ∨	Admini... ∨	aib_revenue - explan... ∨	aib_rev... ∨	aib_revenue... ∨
✓	Customer1	8/19/2020 8:1...	1	0.00	[{"entityName": "a...	No	0.43
	Customer1	8/19/2020 8:1...	1	0.00	[{"entityName": "a...	No	0.43
	Customer10	8/19/2020 8:1...	0	0.00	[{"entityName": "a...	No	0.03
	Customer10	8/19/2020 8:1...	0	0.00	[{"entityName": "a...	No	0.03
	Customer100	8/19/2020 8:1...	3	25.20	---	---	---
	Customer100	8/19/2020 8:1...	3	25.20	---	---	---
	Customer1000	8/19/2020 8:1...	0	0.00	---	---	---

Figure 6-43. *Published view*

Detect Objects

Another supercool thing you have to try out in AI Builder is the object detection model. The model lets you simply recognize common object(s), object(s) in retail shelves, or brand logos present in an image, surrounded by a bounding box like in Figure 6-44.

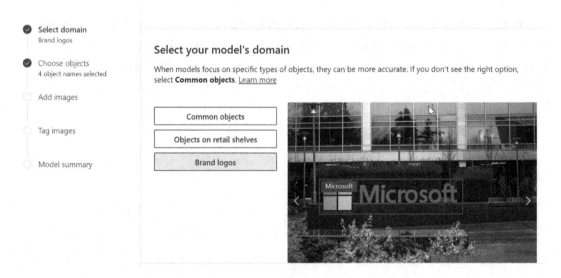

Figure 6-44. *Object detection*

AI Builder asks you to set the name for the object(s) that you would like to detect, and you have to upload and tag at least 15 images per object before the model can be trained. This is really the simplest thing in the world to try out. I promise you! Once you have gone through the manual tagging, trained the model, and published it, you can use it in an app or a flow. In a PowerApp, the object detector will create a scanner for you that lets you upload or take an image to detect like in Figure 6-45. And you can add a data table with count of objects, name of object, time of picture, etc.

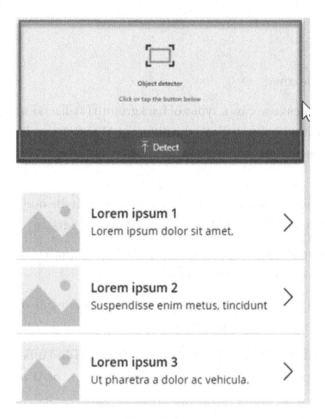

Figure 6-45. *PowerApps with an object detector*

How can this be useful? Imagine in your organization you would like to identify certain brand placements or count brands in a store shelf – to detect if you should refill or to detect parts in a manufacturing facility because you would like to get the manual for a specific part. Another relevant thing now in the time of COVID is to count employees in the office or to see if there are above 20 persons in a crowd. Well, the latter may have some ethical concerns related to it – but imagine all the use cases where this can apply!

Note   This is a preview functionality at the time of writing, meaning it has not reached the production environment and is not intended to be used before it's an official release from Microsoft.

More information on how to build and customize models can be found here: `https://docs.microsoft.com/en-us/ai-builder/model-types`.

Summary

In this chapter you, learned

- What business use cases, types of background (skillsets), and prerequisites (technological) the different AI options on Power Platform need and will help you with when creating your own power solutions

- How to build and deploy a solution powered with the built-in AI Insights in both Power BI Desktop and Power BI Service

- Building and deploying a Power BI AI solution using a custom function calling Cognitive Services APIs in Azure

- How the AI capabilities can bring additional value while developing a solution in terms of data exploration

- How end users of the solutions can gain valuable and unknown insights from data through use of the different type of AI visuals in Power BI

- How to build powerful apps and/or automation processes with AI Builder with prebuilt models and typical use cases for custom models

To learn more, visit `https://powerbi.microsoft.com/en-us/blog/` or Microsoft learning pages.

We hope that you have enough information to understand what the AI in Power BI and Power Platform options are and are ready to go ahead and test for your own use cases.

CHAPTER 7

Chatbots

Chatbots provide a way to build a conversation application between a human and an application using a defined data source. Chatbots are often used in web-based customer service areas to answer common questions as they can simulate human communication and interact intelligently through a dialog with a human. This kind of interaction works best where the questions are likely targeted to a specific topic, for example, restaurant reservations or standard ordering-related questions. When chatbots work well, people interacting with them may not even realize they are not talking to a person.

Chatbot services can be used to create automated routine workflows. These workflows can be implemented to assist customer service by having the chatbot automatically answer simple common questions and refer more complicated questions to customer representatives. A chatbot could easily answer common questions such as "When will my order ship?" or "Can I return this item?" The chatbot interface even provides an alternative way to assist people who may have issues communicating in different ways. People with hearing issues or who are in a noisy environment may prefer to chat as that method of communication may work better. It is also possible to add chatbots faster than adding more staff, allowing customers to get an answer more quickly than waiting for a person to answer a call. Well-designed chatbots are able to use natural language constructs allowing people to use conversational terms that can include jargon known to some groups of people or refer to elements by names used more commonly. People can use their knowledge and user experience of sending SMS on their phone, applying to this service.

This chapter describes how to create a simple chatbot service using the QnA Maker service. This method introduces the Language Understanding (LUIS) service to break down the conversation into different components, which are classified as intents and actions. After that, you will learn how to use the Bot Framework SDK to build up the conversation that includes the business logic behind it.

© Alicia Moniz, Matt Gordon, Ida Bergum, Mia Chang, Ginger Grant 2021
A. Moniz et al., *Beginning Azure Cognitive Services*, https://doi.org/10.1007/978-1-4842-7176-6_7

The chatbot in this chapter is created using Python. The examples will be performed using Azure Portal and macOS using Python 3.8 in Visual Studio Code. Before starting, please make sure you are familiar with the concepts in Chapter 4, especially as it relates to language.

Chatbot Overview

Eliza is considered the first chatbot service, and it was created in 1964 to demonstrate how people and machines could communicate. Eliza was designed to work as if a person was talking to their therapist [3]. Chatbots grew out of this research and used it to rank complexity in conversation. There are commonly thought to be five different levels of conversations. These are based on the complexity of the question that a chatbot can assist the user with, and these levels of understanding are used to create bots of different levels of complexity. The five bot types are notification assistants, FAQ assistants, contextual assistants, personal assistants, and autonomously organized assistants.

Notification Assistants

Notification assistants are designed to notify when requirements are met. For example, once a task has been marked as complete, send a notification to a manager. For notification assistants, once the rules have been created, there are no interactions required. Notification assistants are the simplest kind of bot as there is no interaction and the communication is only one way. No responses are expected nor processed.

FAQ Assistants

Interaction between people and applications is introduced in FAQ assistants. Someone enters a question, and the chatbot will respond with the best answer in its dataset. FAQ assistants are still rather simple as they can handle only single-sentence questions and they do not store any information regarding questions asked earlier, so they are unable to include or act upon any context drawn from previous questions.

Contextual Assistants

The contextual assistant provides a more humanized and consistent experience as it includes interfaces to the past and other services. The chatbot collects the user historical conversation logs and extracts the meaning of the conversation. Instead of a single sentence, the contextual assistant has the ability to understand the meaning of multiple sentences. A contextual assistant is also able to call other services to accomplish user requests. Contextual assistants are limited in that they will not remember a customer's name and their personal data, as such services are not designed to provide personalized service, but instead deliver a consistent experience for all customers.

Personal Assistants

Personalized context is available within the personal assistant style of bot. A good personal assistant behaves like an executive secretary, like JARVIS in the *Ironman* comics. The assistant should be able to deal with different types of tasks such as booking a meeting, booking reservations for a planned trip, and managing your food delivery by knowing your food preferences. This type of assistant uses stored information from prior purchases, your appointment history, and other personal preference data to be able to complete its tasks. Personalized assistants are able to understand conversations, keep the preferences of the customer, provide recommendations, and provide integrated responses. The personalized experience usually brings greater user satisfaction and higher loyalty to the service. This level of performance chatbot is significantly more complex and requires considerably more effort than the previously described conversation types.

Autonomously Organized Assistants

Autonomously organized assistants can learn from user data dynamically and are able to adjust their response according to the latest user input. Autonomously organized assistants run without human supervision. They can handle tasks across multiple areas, releasing users from time-consuming manual tasks.

There are not many systems that can be called autonomously organized assistants as these are quite complex systems to create. They are mainly aspirational and are found sometimes to handle smaller tasks where heuristics and machine learning insight can be used to make decisions.

Autonomously organized assistants may pull data from a number of different areas including current events to provide decision making. This can include analyzing previously identified trends, which may be created through machine learning analysis of the data.

Levels of Understanding Progression

After reviewing these five levels, you will see the progression from using simple text to interact with a chatbot system to ending up with interaction previously conceived in comic books. Chatbots can now be very sophisticated as they can now understand more complex text with context spread across the entire conversation and business history. Interaction can include starting with audio instruction to trigger the chatbot, as seen in devices such as Amazon's Alexa to interact. Chatbots have become a normal means that businesses use to converse with people. Using chatbot interfaces allows customer conversations to be logged and analyzed easily. The data gathered from these conversations reveals more information for business decision making and to be used in further refining the chatbots or perhaps changing the way they do business.

Chatbot Application Use Cases

There are many advantages that companies receive by using chatbots. They can substitute for trained staff and do not care what hours they work. They can supplement existing processes or create new areas to interact with people where they could not devote resources due to business constraints. According to Facebook, there are only 300,000 chatbots among six million advertisers. Not many companies know how to use this technology to improve their business. However, the market is growing as companies start seeing the benefits of another channel of customer interaction. For example, many educational institutions are looking at integrating chatbots into their environments to improve student interaction in online courses.

As much communication now is done via phone text message, customers feel comfortable texting companies as this is how they engage in conversation. Unlike many other customer service interactions, companies do not need to change business logic to integrate a chatbot. You can easily plug in the service and test if it works with your business and tune it to fit to the customer experience flow. We can interact with a chatbot as another form of human-to-computer interface. There are many possible ways to deliver positive experiences through chatbot conversations. Because chatbots are available 24/7, they provide higher frequency of communication between business and customers. More frequent communication can provide a better experience for resolving customer problems, as they will be able to contact a part of the company at all hours as chatbots never sleep.

There are a number of different chatbots in use throughout different industries as most are looking to incorporate chatbots if they have not already done so. Banking, healthcare, education, professional services, and marketing organizations have all developed chatbots to respond to people requesting information or assistance.

Chatbot Terms and Definitions

There is a different language used to describe the different elements included in a chatbot. In this section, we will provide examples of what the different terms mean and how to use them.

Domain Words

There are three different domain words: *utterance*, *intent*, and *entity*. These three words are used to describe the subject matter of the chatbot.

An utterance is anything the user says. For example, if a user says, "I would like to book a ticket to London on next Friday," the entire sentence is the utterance.

The intent is the verb or desired action along with the object of the action. The intent is what a user is trying to accomplish. For example, if a user says, "I would like to book a ticket to London on next Friday," the user's intent is to book a ticket. Intents are given a name, often a verb and an object, such as "book tickets."

An entity modifies an intent by adding specificity. For example, if a user says, "I would like to book a ticket to London on next Friday," here the entity is the timeframe of Friday, which modifies the previous intent. Entities are given a name, such as "location" and "dateTime." Entities are sometimes referred to as *slots*.

Chatbot Scope and Definition

When creating a chatbot, the first step is to establish the chatbot's goal. What is it that the chatbot is supposed to accomplish? This needs to be clear in the mind of the developer and the requesting organization to make sure the chatbot's tasks are clearly defined. Before creating a dialog, meet with the requesting organization to find out more about the people who will be interacting with the chatbot. To develop the chatbot, you will need to know the answers to a number of different questions, including these:

- How would you describe the target audience?

- What, if any, kind of personality should the chatbot have?

- Should the chatbot have a sense of humor, or should it have purely a professional tone?

- What kind of information is important to know prior to beginning development?

The communication flow between chatbot and person needs to be defined. When should the chatbot move to contact a person? What are the key words or phrases that will indicate a response? How will the chatbot respond to profanity? These questions need to be part of the design. Planning should also occur on how to update the chatbot's flow based upon real-world data received when it is in use. New words and phrases will need to be added to the chatbot's repertoire of responses. The development process needs to include updates and processes needed throughout the chatbot lifecycle.

Development Tips

Before getting started, make sure that you understand the different scenarios the chatbot is designed to solve, paying extra attention to those that have been identified as the most important. Identify the conditions that might need human intervention and when you need to update information the chatbot uses to respond. For example, the chatbot service can understand the conversation in most cases, but sometimes there will be phrases that are not included in the domain databases, which might lead to error messages or logic loops. Try to anticipate when a possible loop can occur in order to prevent those scenarios from happening. Improving the chatbot service is an important part of the design. Make sure there is a place that you can update the responses or update the actions performed in response to the real-world use. Another condition that needs to be resolved is if there is a step that leads the user to an unfinished conversation where the chatbot says nothing or ends the conversation. Whenever possible, this condition needs to be eliminated to ensure customers are guided to where they can receive further information.

Guidelines for Building a Chatbot

When designing a chatbot system, there are several aspects you will need to consider. These include strategy, design considerations, channels (i.e., user interface), AI capabilities, source systems, and system-level setup. Figure 7-1 provides a diagram showing how these different aspects are incorporated into a chatbot design.

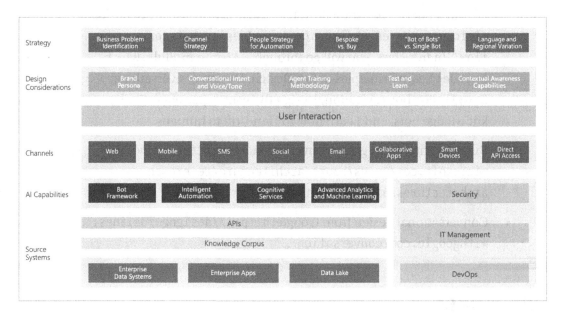

Figure 7-1. *Five components for developing chatbots*

Also important are the following elements that need to be defined prior to beginning development:

- Bot logic and user experience

- Bot cognition and intelligence

- Data ingestion

- Logging and monitoring

- Security and governance

- Quality assurance and enhancements

- Design considerations

Chatbot construction needs to incorporate existing security practices and ideally include DevOps for optimal development. Chatbot development should start with the business challenges to resolve as these should drive the main user story and user experience flow. Once these elements are known, you have the information needed to determine which AI service and framework to select to best fit your design use cases.

In addition to the guidance we've just given, here are three resources that can help further your understanding of conversational design:

- Design a bot conversation flow, from Microsoft Learn: `https://docs.microsoft.com/en-us/learn/modules/design-bot-conversation-flow/`. In this class, you will go through a 32-min module, which includes Learn Principles of Bot Design, Design conversational flow, Design bot navigation, Design the user experience, Design knowledge bots, and Learn how to Hand-off to humans.

- Build an enterprise-grade conversational bot: `https://docs.microsoft.com/en-us/azure/architecture/reference-architectures/ai/conversational-bot`.

- Conversational design from Google: `https://designguidelines.withgoogle.com/conversation/`.

QnA Maker

QnA Maker is a cloud-based API service from Microsoft Azure that can be used as part of your conversational application service. QnA Maker can be helpful if you are designing a bot to respond to frequently asked questions from a specific source such as a user manual.

Documentation is extracted to become the knowledge base used by the chatbot. You can update the knowledge base to the latest version with the latest information and updates to the documentation.

Since QnA Maker is an API service, you can use the REST interface to interact with the system. The service can be used to edit content or train the conversation.

Examine the QnA Maker website API page for more detailed information on how to train it. It is also possible to add some personality to this QnA service with responses that can be more than just an answer. All the deployment and data contained within QnA Maker are secure and privacy compliant.

QnA Maker would be the best tool for developing chatbots when the following conditions occur:

- When you have static information as your reply answers

- When you expect to have the same response for all the users

- When you have a knowledge document as your response reference

Developing a Knowledge-Based Bot

QnA Maker is dependent upon having a knowledge base containing the data responses incorporated in its development. The knowledge base will need to be maintained to ensure it has the latest information and includes information that reflects questions users are asking. Figure 7-2 shows the ongoing development lifecycle of the knowledge base used with the bot.

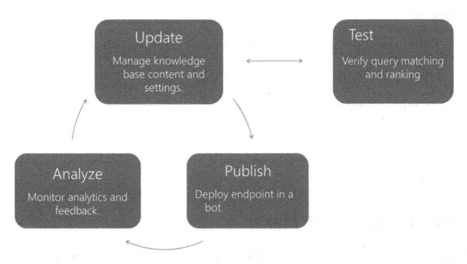

Figure 7-2. *The lifecycle of the knowledge base [7]*

Test and Update

The knowledge base is the most important part of the service. Use the Test function to perform tests with common user input and verify if the response is correct. In this phase, try many times and try with different entries till you are satisfied with the response the chatbot is providing. The testing you do here will not affect the production environment. If you would like to test the service with the QnA Maker API, use the istest setting to specify which testing environment to use.

Publish

After testing, it is time to publish you application to production. Publishing will send the latest knowledge base to the Azure Cognitive Search. Publishing also updates the service endpoint.

Analyze

To analyze the QnA service, you will need to enable diagnostics logging in the QnA Maker managed service. You can select which metrics you would like to monitor, including Audit, Request and Response, and Trace. The common scenarios monitored for the logs are

- Traffic count per knowledge base and user in a time period

- Latency of the Generate Answer API

- Average latency of all operations

- Unanswered questions

Having this information in the logs will assist in the analysis to further improve the performance over time.

Estimated Costs for QnA Maker

QnA Maker is not a standalone service. It will need to be incorporated with Azure App Service and with Azure Cognitive Search for searching the response data. You might also want Application Insights, which is an optional monitoring service. You will need to look at the price page for Azure App Service, Azure Cognitive Search, and Application Insights to be able to determine the cost of the entire solution.

Each service on Microsoft Azure has its own price page. The following example pricing is based on the Central US data center:

- *Free tier* – Three transactions per second with a maximum of 100 transactions per minute and maximum of 50,000 transactions per month. Storage for a document of no more than 1 MB and three managed documents per month.

- *Standard tier* – Three transactions per second with a maximum of 100 transactions per minute. Documents are $10 a month for unlimited managed documents. A managed document can be a URL or page in .tsv, .pdf, .doc, .docx, .xls, or .xlsx format.

Building a Simple QnA Service

This section shows the steps required for creating a QnA service.

Step 1 – Go to `www.qnamaker.ai/`, and click the **Get started** text on the web page. A picture of the website is shown in Figure 7-3.

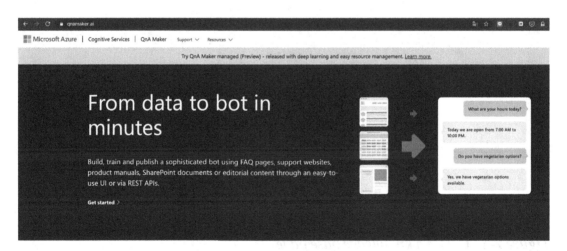

Figure 7-3. *QnA Maker service page*

Step 2 – After the first step, you will be directed to a page where you can create a knowledge base. To do this, click the **Create a knowledge base** text as shown in Figure 7-4.

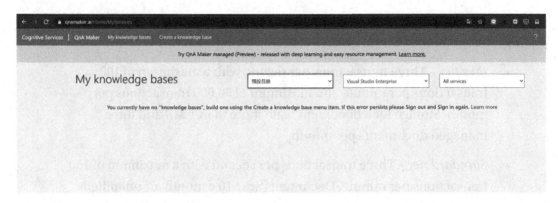

Figure 7-4. *My knowledge bases page*

Step 3 – A new page will load containing a form you will need to complete to create the knowledge base in Azure. A picture of the form is shown in Figure 7-5.

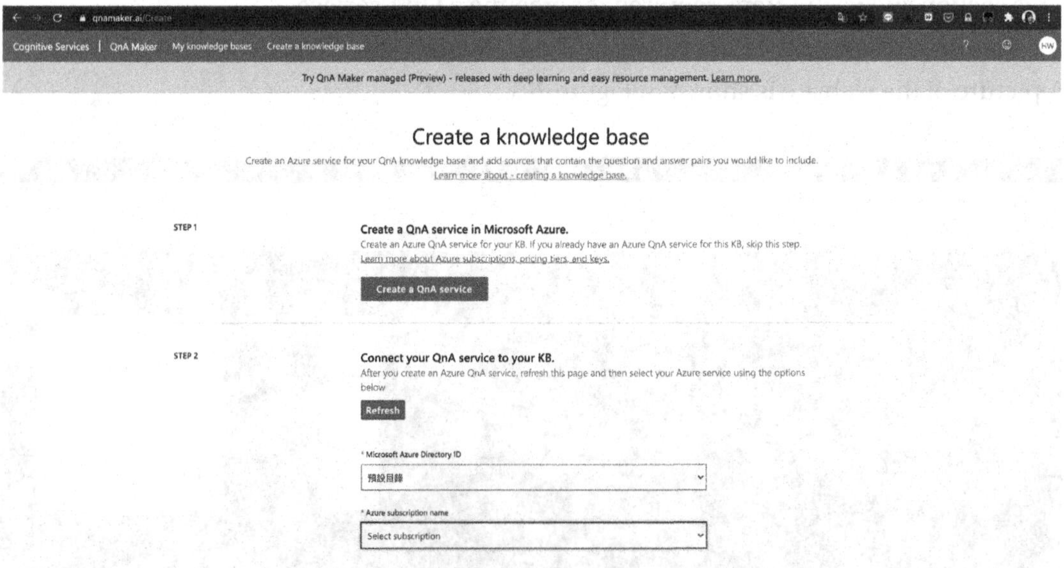

Figure 7-5. *Create a knowledge base page*

You will need to click the button labeled **Create a QnA service**, which is shown on the form.

Step 4 – After clicking the **Create a QnA service** button, the screen will change allowing you to select several options from dropdown boxes, including the service name, pricing tier, resource group, and region. A picture of the form is included in Figure 7-6, which is shown here.

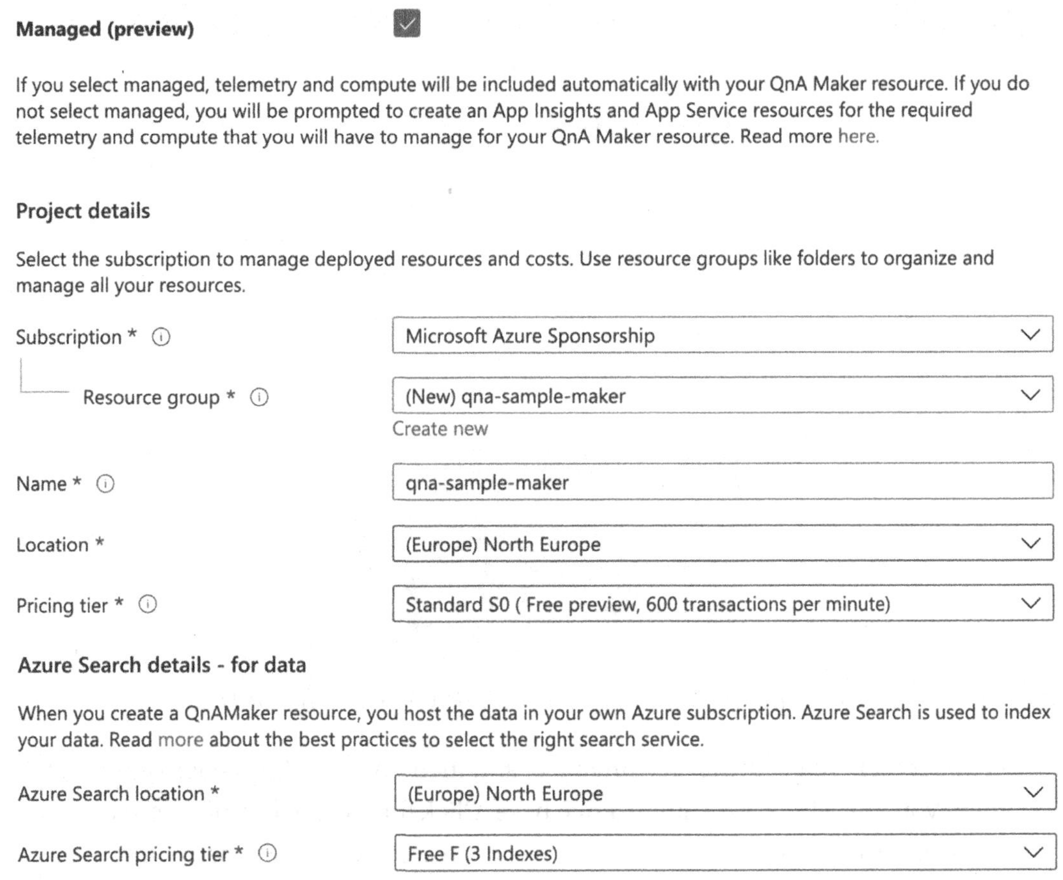

Managed (preview)   ☑

If you select managed, telemetry and compute will be included automatically with your QnA Maker resource. If you do not select managed, you will be prompted to create an App Insights and App Service resources for the required telemetry and compute that you will have to manage for your QnA Maker resource. Read more here.

Project details

Select the subscription to manage deployed resources and costs. Use resource groups like folders to organize and manage all your resources.

Subscription * ⓘ	Microsoft Azure Sponsorship ⌄
Resource group * ⓘ	(New) qna-sample-maker ⌄
	Create new
Name * ⓘ	qna-sample-maker
Location *	(Europe) North Europe ⌄
Pricing tier * ⓘ	Standard S0 (Free preview, 600 transactions per minute) ⌄

Azure Search details - for data

When you create a QnAMaker resource, you host the data in your own Azure subscription. Azure Search is used to index your data. Read more about the best practices to select the right search service.

Azure Search location *	(Europe) North Europe ⌄
Azure Search pricing tier * ⓘ	Free F (3 Indexes) ⌄

Figure 7-6. *Detail setup for the QnA Maker service*

After selecting your desired answers from the dropdowns, click the Create button to publish the service. This process usually takes a few minutes to complete, and when it does, a congratulation page will appear.

Step 5 – After the QnA service is complete, you will need to go back to the page containing the button for creating a QnA service so that you can create a knowledge base for the service. A picture of this screen is shown in Figure 7-7.

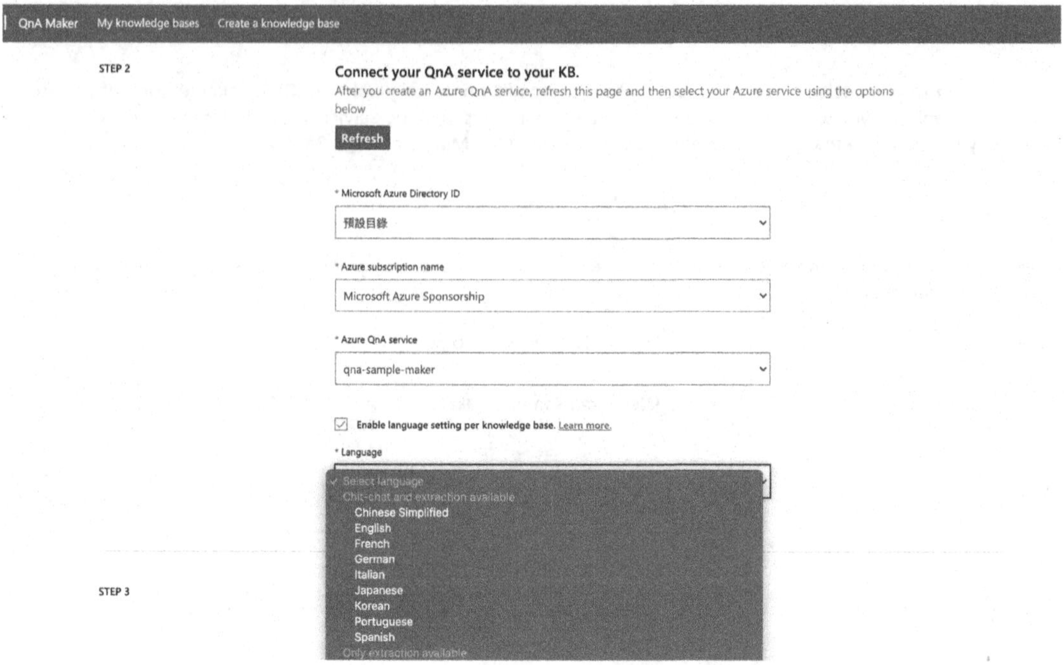

Figure 7-7. *Connect the QnA Maker service to the knowledge base*

On this screen, select the target language, and then link your Q&A page or upload the document you are using for the knowledge base. Click the **Create** button to create your knowledge base. Once the knowledge base has been created, you will see two sections: your document and the chatbot personality documents.

Step 6 – After clicking the document, you can see a preview of the knowledge base where you will need to update the questions and answers, which is shown in Figure 7-8.

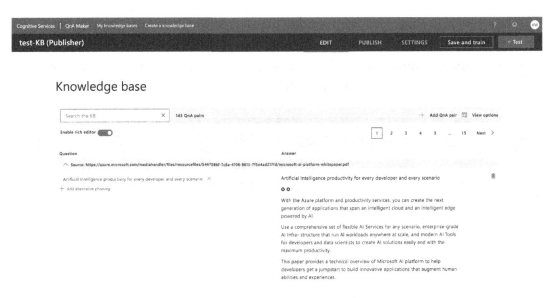

Figure 7-8. *Review the knowledge base content*

After all of the questions and answers are completed, you can start training your model.

Step 7 – After the training is complete, the chatbot is ready for testing as shown. Tests will incorporate the information from the previously added knowledge base. To test the chatbot, click the blue button labeled **Test**, which can be seen in Figure 7-9.

Figure 7-9. *Testing QnA Maker settings*

The screen will show the responses that the QnA Maker is using in the chatbot.

Step 8 – When you have completed testing, click the button labeled **Publish**, which is to the left of the **Test** button on the same page. The screen will change, and you will get a message to confirm as shown in Figure 7-10.

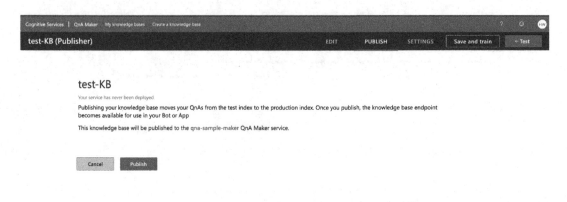

Figure 7-10. *Publish the QnA Maker service confirmation*

When this action is complete, you will receive a success message and a script that can use Postman or Curl to interact with the endpoint. A picture of this Success screen is shown in Figure 7-11.

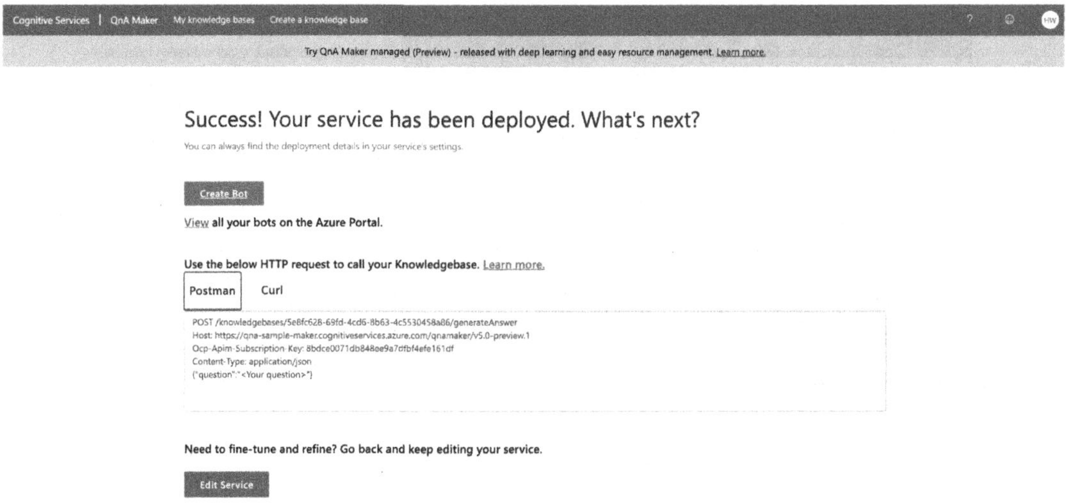

Figure 7-11. *QnA Maker service deployment Success screen*

Step 9 – The next step is to complete the form for creating a Web App Bot. You will need to go to Microsoft Azure Portal and search for the Web App Bot service. Once you have selected this service, you will be shown the form displayed in Figure 7-12.

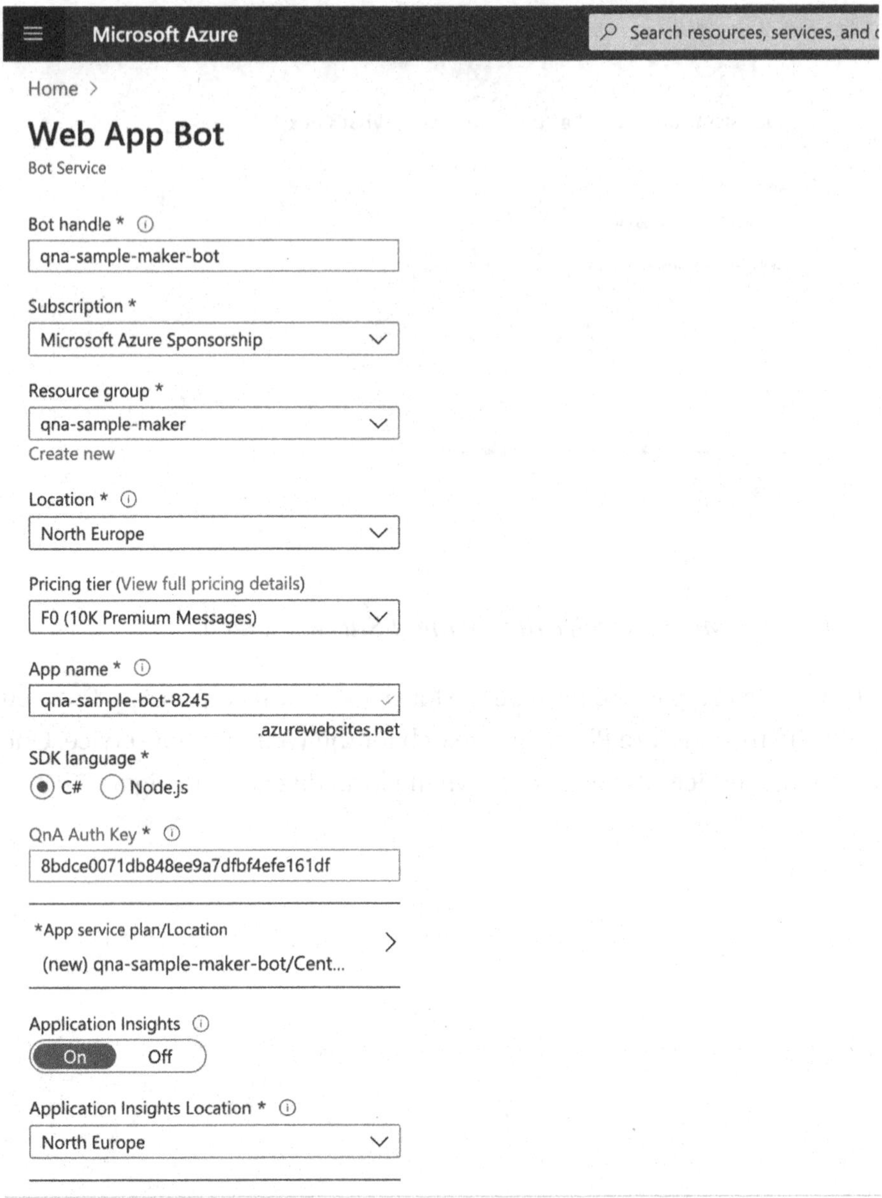

Figure 7-12. *Web App Bot*

Complete the blank elements on the form, including name, location, and pricing tier. When you have completed the form, you can create the bot service.

Step 10 – On the chatbot page, you can click "Test in Web Chat" to access the web test interface, which you can see in Figure 7-13.

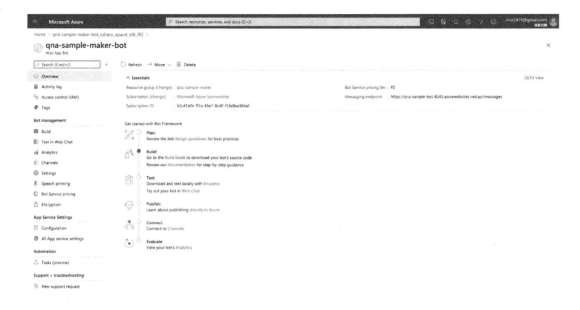

Figure 7-13. *Web App Bot Page*

After selecting the **Test in Web Chat** option from the menu on the left-hand side of the screen, you may access the web test interface.

Step 11 – To test the chatbot in the web chat, use the form shown in Figure 7-14 to test questions and responses.

Figure 7-14. *Web chat test interface*

If you identify any responses that should be improved, go back to the training page shown in Figure 7-9, which we reviewed earlier. Start another test run where you edit the response and retrain the model.

Step 12 – Using the menu on the left side of the screen, select the option Channels. A screen will then appear as shown in Figure 7-15.

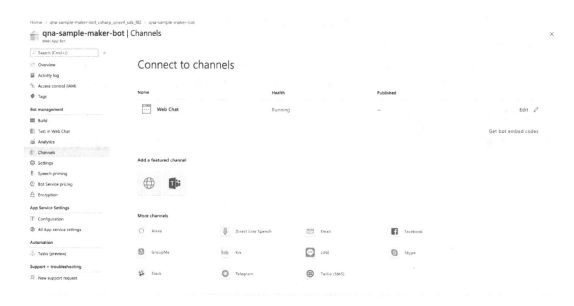

Figure 7-15. *Connect to channels page*

Notice on the screen there are many different channel options. These different options provide methods you can use to interact with your customers, including Alexa and Telegram.

Step 13 – Once you select a channel from the list, the screen will change appearing like Figure 7-16.

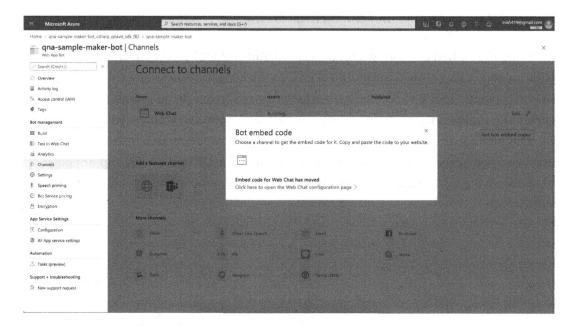

Figure 7-16. *Get chatbot embed code message box*

To get the embed code for the channel you selected, select **Click here to open the Web Chat configuration page**.

Step 14 – The next screen, Configure Web Chat, as shown in Figure 7-17, provides the secret keys needed to configure the application to use the chatbot as well as the embed code needed for your application.

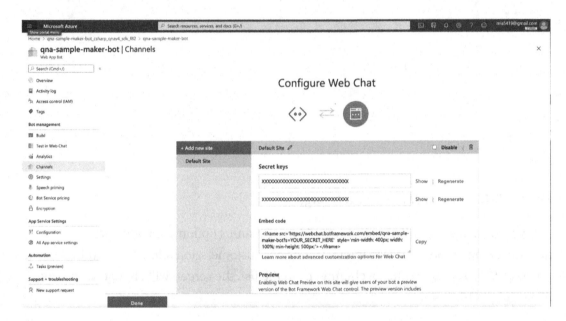

Figure 7-17. *Secret keys and embed code for the QnA chatbot*

Copy the embed code shown on the screen and replace "YOUR_SECRET_HERE" with the secret keys shown in Figure 7-17. Click the Show button to make the keys visible before copying them. To add the chatbot to your website, create an iframe and paste the code as an iframe object in your website. You have a chatbot in your website.

Optional Customization Step – Review the source code generated in the UI. If you are interested in seeing the code that is developed by the UI, in the Web App Bot page as shown in Figure 7-18, you can click **Build** on the left menu bar to download the bot source code.

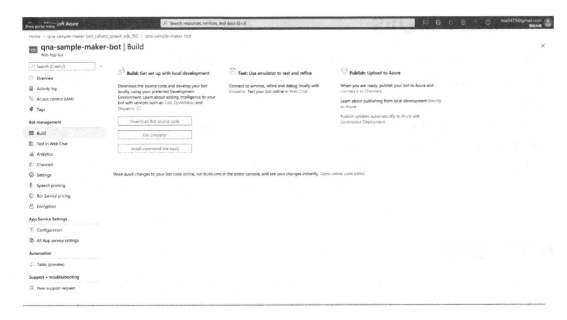

Figure 7-18. *QnA chatbot code download page*

QnA Maker Summary

The steps outlined in this section have shown how to create a QnA Maker service using the user interface to select and implement the required components. We also examined some elements that you need to consider when setting up a QnA Maker service. The steps also demonstrated that the service can be tested in an iframe object that you can easily use in your website.

Selecting QnA Maker for your chatbot is a good choice when you have static content, including documents, manuals, or white papers where you want to create an identical user experience across all customers. If you have a more complicated conversation, which may contain a bigger set of knowledge bases, or you need different experiences for different classifications of customers, QnA Maker is not the best choice. For those conditions, the service you want to use in your chatbot is Language Understanding Intelligent Service (LUIS).

Introduction to LUIS

LUIS, which stands for Language Understanding Intelligent Service, is a cloud-based Microsoft AI service. It can predict the intention of a conversation and extract the key information. LUIS is designed to integrate with the Azure Text Analytics and Speech Analytics services and with Azure Bot Service, to provide a comprehensive conversational service experience.

There are three Azure resources used when provisioning a LUIS service. These are the authoring resource, the prediction resource, and the Cognitive Services multiservice resource.

Authoring Resource

The authoring resource allows you to create, manage, train, test, and publish your LUIS applications. Unfortunately, this resource is not available in all Azure data centers, so you will want to check to see where it is available. To create a LUIS app programmatically or from the portal, you will need to link the authoring resource to the application. It does come with a free tier (F0) that you can use for development. The free tier provides one million authoring transactions and 1,000 testing prediction endpoint requests per month.

Prediction Resource

The prediction resource implements the prediction service that is part of LUIS. For the free tier (F0), you may use 10,000 predictions for text input per month. If you choose to implement the standard tier (S0), additional functionality is available as that tier includes text and speech prediction features.

Cognitive Services Multiservice Resource

When using Azure Cognitive Services, you have the option of using a single service or multiple services. Developing a chatbot requires more than a single service, so you need to create a multiservice subscription key. This will provide the ability to use LUIS and other integrated Cognitive Services in an endpoint. The multiservice resource key is used to programmatically manage multiple keys.

To manage the security for these objects, you will be using Azure's Role-Based Access Control (RBAC). These objects need to be added to users using standard RBAC methods for access.

LUIS Conversation Processes

LUIS processes conversations using the design pattern for text entered into the chatbot. LUIS follows the pattern of receiving, processing, and then responding.

For example, imagine that the text "I would like to book a train ticket to London on next Friday" is sent to the LUIS REST API endpoint. LUIS processes the text using the natural language processing (NPL) model to identify the meaning of the sentence. That model provides a score or percentage predicting how accurate it believes it is when determining the meaning of the text. This prediction is used to select a response, which LUIS sends to the customer in a JSON format as an output response. The customer uses the response as part of their decision-making process to perhaps add a follow-on question, and the process repeats. These decisions can be included in another bot framework or intelligent application. User input is processed to improve the LUIS service over time.

There are a few rules that can limit an interaction, including rules around text length and dispatch. For text requests, the API transaction limits the input query text length to 500 characters maximum. For speech requests, a transaction is an utterance with query length up to 15 seconds long. Dispatch will do two text transactions per request. Dispatch is a feature that enables processing two models/applications with one API call.

LUIS Creation Tutorial

Here we are going to develop a conversational chatbot. The first step in the design process is to specify the terms used and provide data for them.

Step 1 – Sign into the LUIS portal: `www.luis.ai/`. A picture of the page is included in Figure 7-19.

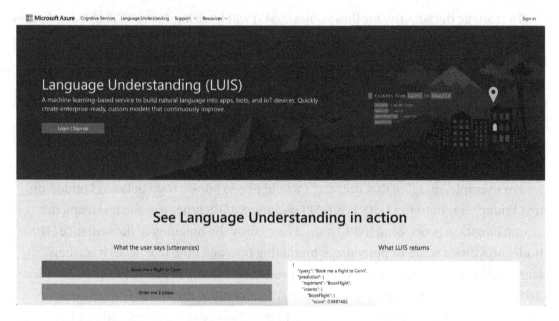

Figure 7-19. *LUIS service page*

Step 2 – Select your subscription from the dropdown list and then create a new
Cognitive Services account resource as shown in Figure 7-20.

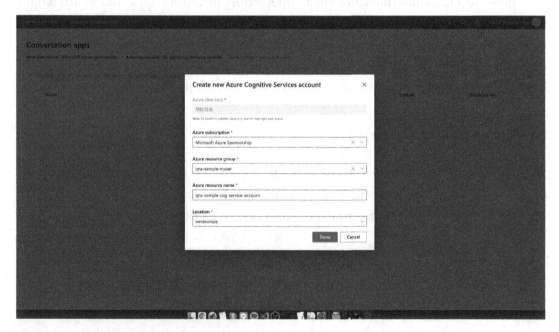

Figure 7-20. *Create a new Azure Cognitive Services account*

Step 3 – After the cognitive services account is created, create a new LUIS app in the LUIS portal as shown in Figure 7-21.

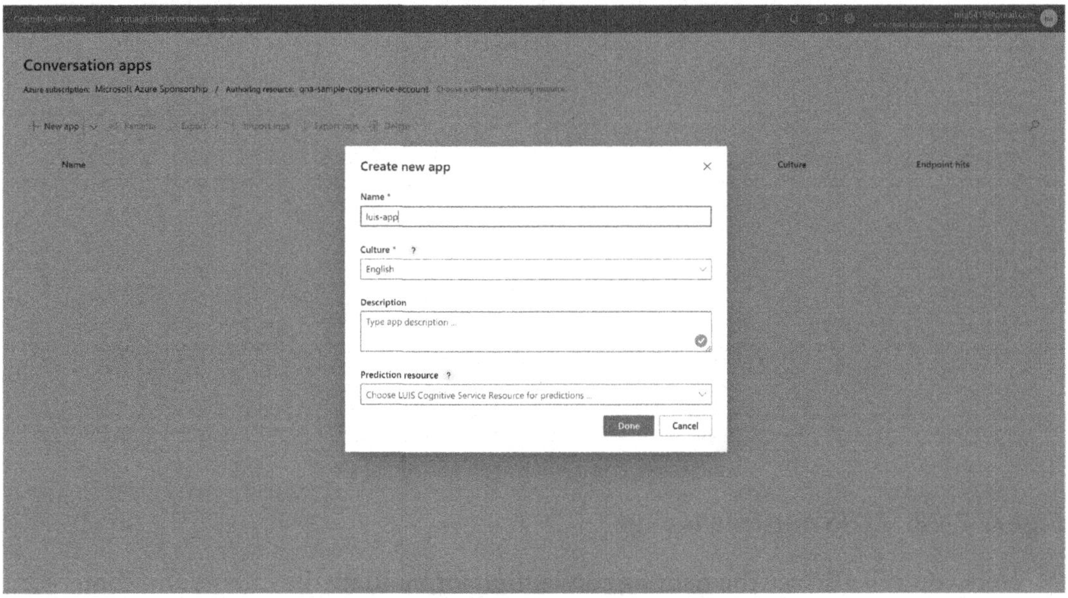

Figure 7-21. *LUIS create new app page*

In the Prediction resource dropdown, select **Choose LUIS Cognitive Service Resource for predictions...** After creating the LUIS app, the browser will show you the welcome page of the LUIS app where you may wish to review the welcome tutorial. The tutorial includes some common use case examples. Close the window when you are done looking at the tutorial.

Step 4 – Add the calendar prebuilt domain intent to the intent list to start the first app as shown in Figure 7-22.

Figure 7-22. *LUIS app intent page*

Here you will also see the naming convention for building the intents and some example intents. The examples can be a helpful way to learn how to identify the intents for use cases and how to prepare the examples when building the intent.

Step 5 – Add Calendar and Utilities from Prebuilt Domains as shown in Figure 7-23.

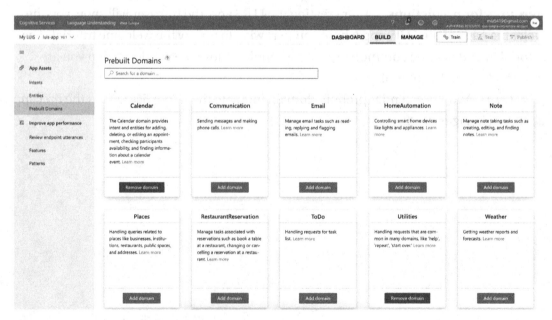

Figure 7-23. *Prebuilt Domains page*

The entities contain the key information that you want to collect for your business process. For example, when you book an appointment, you will need a date, event start time, event end time, location, etc. Most common entities come with the prebuilt domains, but if you can't find them in the prebuilt lists in Figure 7-23, then you will need to bring label data for the entities.

Step 6 – Using the screen shown in Figure 7-24, click the Patterns tab to learn how the LUIS app recognizes your sentences.

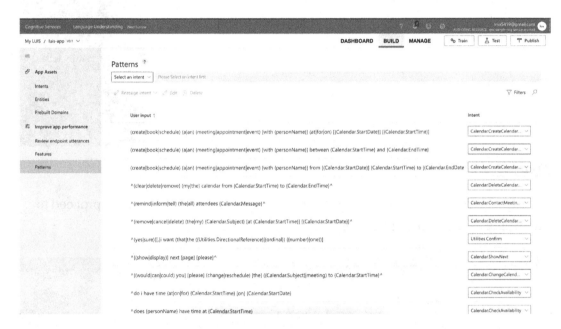

Figure 7-24. *Patterns page in the LUIS app*

Step 7 – When you have all the data prepared, you can start the training process. Click the option **Train** on the upper-right side of the page to start the training job. The screen will then look like it does in Figure 7-25. Wait a few minutes. When the training job is finished, you may start testing.

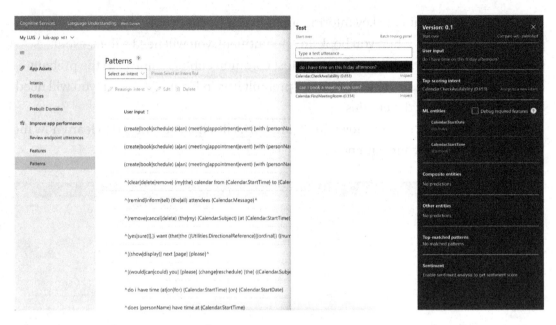

Figure 7-25. *The testing panel*

Step 8 – When you are satisfied with the testing results, you will want to proceed to publishing as shown in Figure 7-26.

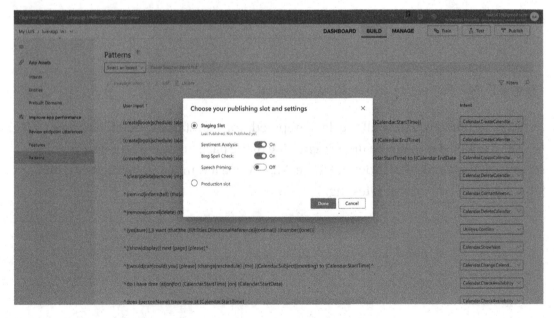

Figure 7-26. *Publish settings*

You will see the popup box Choose your publishing slot and settings. In the setting box, you may select the production or staging environment to deploy. There are also several optional elements to add sentiment analysis, Bing spell-check, and speech priming.

Step 9 – You will be notified after successfully publishing the service and receive a notification for the status as shown in Figure 7-27. The notification messages will be visible when on the notification bell on the upper-right menu bar.

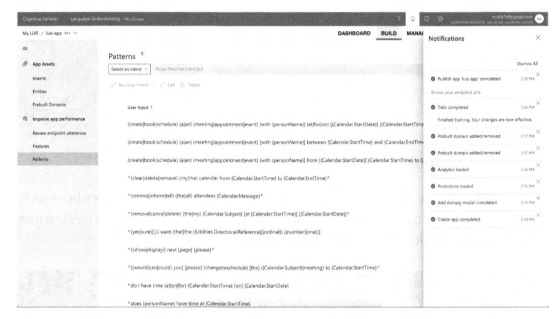

Figure 7-27. *Notification messages*

Step 10 – Once the LUIS app is deployed to the staging environment, you will need to add a production resource. Go to the Manage tab and select **Add prediction resource** as shown in Figure 7-28.

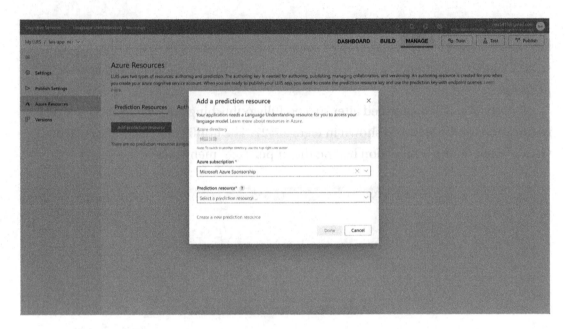

Figure 7-28. *Prediction resource popup*

After selecting the subscription, select **Create a new prediction resource** because at this point you do not have one to select. This will bring up a popup menu to create a new Azure Cognitive Services account as shown in Figure 7-29.

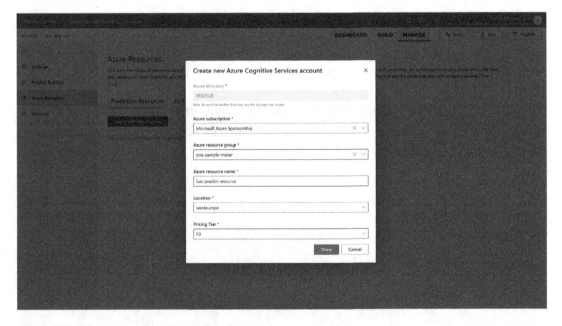

Figure 7-29. *Create a Cognitive Services account*

Connect your deployed service to the new Cognitive Services account created in the popup. Once these steps are complete, you can then create a resource for the service.

Step 11 – The screen shown in Figure 7-30 shows the Azure Resources page you will need for adding a prediction resource.

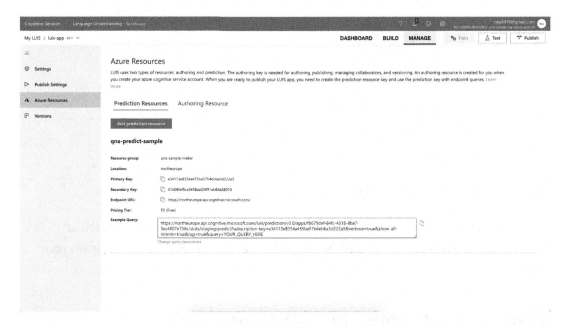

Figure 7-30. *Azure Resources page for prediction resource*

Click the blue button Add prediction resource.

Step 12 – After all of the creation steps are complete, you may want to spend some time testing the LUIS bot. Testing results are available in the Dashboard tab as shown in Figure 7-31.

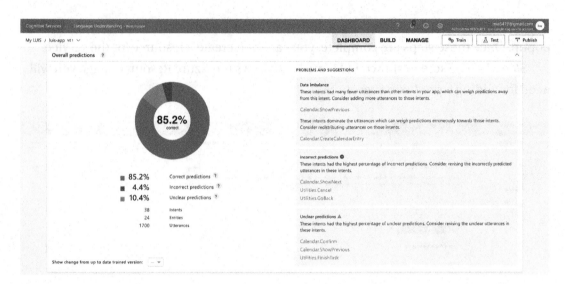

Figure 7-31. *Dashboard of QnA Maker predictions*

The dashboard includes correct prediction, incorrect prediction, and unclear prediction data points. Use the dashboard and check the suggestions to improve your data.

Using Chatbots with QnA Maker and LUIS

Previously we have reviewed using the UI, and now we are to switch to developing via code using Python. We will walk through an example application showing how to run an application using the QnA Maker service and LUIS service, which we created in the previous steps. You will need to complete those steps prior to beginning this section.

The code in Listing 7-1 will be used to create a bot, using Bot Framework Emulator, which is a desktop application for bot developers. It can be installed on Windows, macOS, and Linux. Developers can use it to test and debug the bot app locally with real-time system logs. Or you may run the bots remotely.

Step 1 – Clone the code from GitHub to your local computer so that you can modify the example.

Step 2 – Download Bot Framework Emulator from GitHub: `https://github.com/ Microsoft/BotFramework-Emulator/releases/tag/v4.11.0`. As Figure 7-32 shows, you may download the emulator from the links of assets.

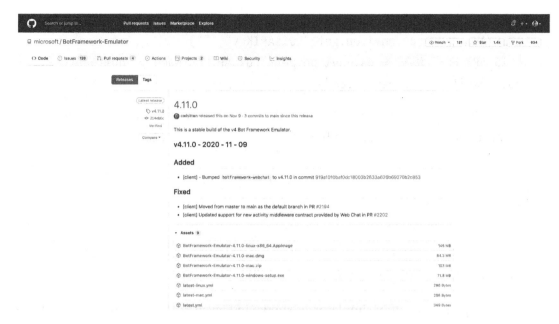

Figure 7-32. *The emulator download page*

Step 3 – Open the project in Visual Studio Code and run `pip3 install -r requirements.txt` in the VS Code terminal. Open the config file, and update your config with your QnA service, LUIS service, app ID, and password.

For the app info and the QnA service info, you may find the information in Azure Portal. For the LUIS service, go to `www.luis.ai/applications`, and find the LUIS app you deployed earlier. The LUIS API host name depends on the region for the created resources.

Listing 7-1. Config for Bot

```
class DefaultConfig:
    """ Bot Configuration """

    PORT = 3978
    APP_ID = os.environ.get("MicrosoftAppId", "")
    APP_PASSWORD = os.environ.get("MicrosoftAppPassword", "")

    QNA_KNOWLEDGEBASE_ID = os.environ.get("QnAKnowledgebaseId", "")
    QNA_ENDPOINT_KEY = os.environ.get("QnAEndpointKey", "")
    QNA_ENDPOINT_HOST = os.environ.get("QnAEndpointHostName", "")
```

```
LUIS_APP_ID = os.environ.get("LuisAppId", "")
LUIS_API_KEY = os.environ.get("LuisAPIKey", "")
LUIS_API_HOST_NAME = os.environ.get("LuisAPIHostName", "northeurope.
api.cognitive.microsoft.com")
```

Step 4 – In terminal services as shown in Figure 7-33, save and run python app.py. You may run the service at your localhost:3978.

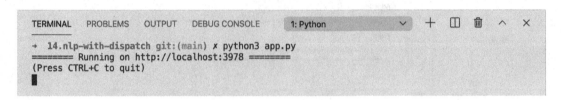

Figure 7-33. *Host your bot application locally*

Step 5 – Open the chatbot simulator, as shown in Figure 7-34, click Open Bot, and enter the host URL, for example, `http://localhost:3978/api/messages`. Next, click Debug Mode, and click Connect.

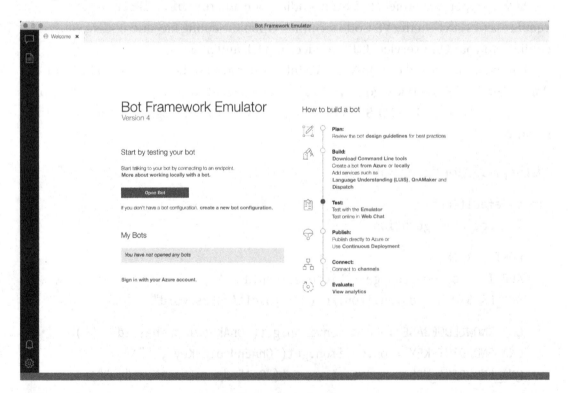

Figure 7-34. *Bot Framework Emulator entry page*

Step 6 – On the left-hand side of the emulator, there is a conversation simulator, and on that side, you can debug as shown in Figure 7-35. Here you can see the message process.

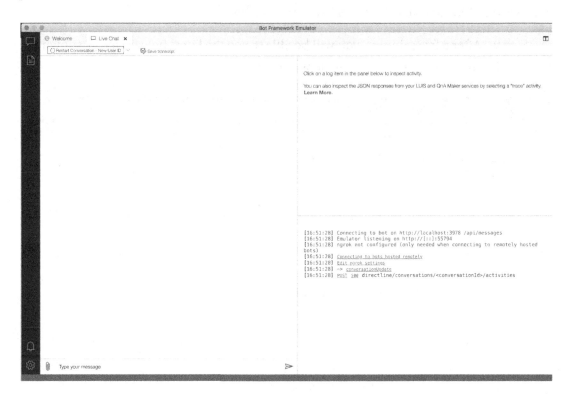

Figure 7-35. *Bot Framework Emulator dialog and message board layout*

Step 7 – The next step is to run the bot in the simulator with or without the debug mode. For more tips about the debug process, see [10] and [11].

Step 8 – Later on if you want to deploy the bot, with an existing service plan, use Listing 7-2 to deploy the service.

Listing 7-2. Deploy the Current Chatbot

```
$ az account set --subscription "<azure-subscription>"
$ az ad app create --display-name "displayName" --password
"AtLeastSixteenCharacters_0" --available-to-other-tenants
$ az deployment group create --resource-group "<name-of-resource-group>"
--template-file "<path-to-template-with-preexisting-rg.json>"
--name "<bot-app-service-name>"
```

```
--parameters appId="<app-id-from-previous-step>" appSecret="<password-from-
previous-step>" botId="<id or bot-app-service-name>" newWebAppName="<bot-
app-service-name>" existingAppServicePlan="<name-of-app-service-plan>"
appServicePlanLocation="<region-location-name>"
```

Step 9 – After a few minutes, you will be able to see the latest bot service as shown in Figure 7-36. You will be able to test it via web chat as well.

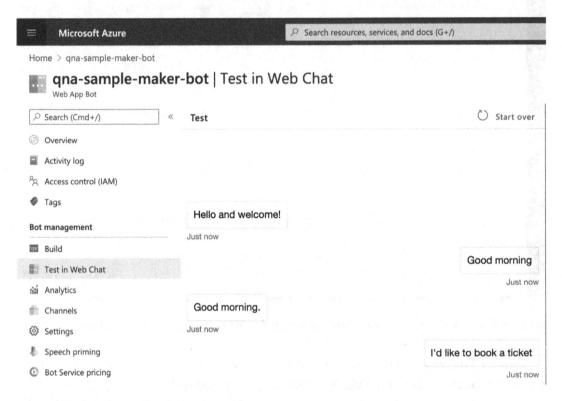

Figure 7-36. *QnA chatbot testing page*

Choosing Between QnA Maker and LUIS

Now that you're familiar with QnA Maker and LUIS, you might be wondering about how to make the best choice between them for whatever application that you are contemplating to create. The following is some general advice that we believe puts you on the right path. Just keep in mind that there may be exceptions and that our advice is general in nature and is best thought of as providing a starting point to your decision making.

We recommend choosing QnA Maker when there's static content that you would like to provide to all the customers. Types of static content can include weather info, the price of a movie ticket, or specific answers to a list of specific questions that you predefine. You can also use QnA Maker whenever you have static information that you wish to share with customers and you want to make it easy for your customers to find the bits of information they need.

Chose LUIS when you want to allow customers the option to speak or type more complex requests that need to be processed by understanding their intention. LUIS brings more Language Understanding capability, which helps you to identify intentions and key information in these more complex conversations.

Summary

Chatbots are very popular as they are being widely adopted by many companies, and here we showed how to create one. This chapter covered the different levels of conversational service and provided a description and steps for creating a QnA Maker service. The LUIS service was also reviewed, and steps were provided to use it. This chapter also covered the different use cases for QnA Maker and LUIS to provide you the information needed to determine which service to use. You learned when the tools should be selected based upon your design goals and how to combine the functionality of both in Python and how to debug and deploy the solution in Azure.

References

[1] www.avanade.com/-/media/asset/thinking/conversational-ai-pov.pdf

[2] https://docs.microsoft.com/en-us/learn/modules/design-bot-conversation-flow/

[3] https://en.wikipedia.org/wiki/ELIZA

[4] https://docs.microsoft.com/en-us/azure/architecture/reference-architectures/ai/conversational-bot

[5] https://designguidelines.withgoogle.com/conversation/

[6] https://docs.microsoft.com/en-us/azure/cognitive-services/luis/luis-reference-regions

[7] https://docs.microsoft.com/en-us/azure/cognitive-services/qnamaker/
concepts/development-lifecycle-knowledge-base

[8] https://azure.microsoft.com/en-gb/resources/microsoft-ai-platform-
whitepaper/

[9] www.cdc.gov/coronavirus/2019-ncov/faq.html

[10] https://docs.microsoft.com/en-us/azure/bot-service/bot-service-debug-
emulator?view=azure-bot-service-4.0&tabs=python

[11] https://docs.microsoft.com/en-us/azure/bot-service/bot-service-debug-
bot?view=azure-bot-service-4.0&tabs=python

CHAPTER 8

Ethics in AI

What does ethics have to do with AI? Why include this topic in a technical book? These may be two of the questions that might have come to mind when you started reading this chapter. As computers start making decisions that may have been previously made by humans, a number of questions may be raised. How is the result determined? Is it based upon a sample set that includes representative samples from a number of different groups? Are there ethical considerations to be made regarding the selection of the datasets? What was the logic used to select groups? Just a random 70% sample of your data? Is the data reflective of the population at large, or does it contain demographic skews that do not reflect the population of the world around you? A little thing like not paying attention to the data used to train your solution can have a major impact upon the people who are depending on the results of your solution.

In this chapter, we are going to take a look at what considerations need to be made when creating, testing, and evaluating an algorithm to ensure you are not biasing the results in favor of one group over another. We are going to be exploring the following topics:

- Items to consider when developing AI

- Ethically selecting the right test data

- Incorporating bias testing in the development process

- Monitoring ongoing learning after the model is released

- Resolving issues when they are raised

Why Is Ethics in AI an Issue?

Ethics in AI is not just a technical issue. Issues in AI have spread to front pages of news websites. One of the first companies to run into an issue with AI was Netflix. In 2010, Netflix created a contest with a million-dollar prize for creating an algorithm that would

281

© Alicia Moniz, Matt Gordon, Ida Bergum, Mia Chang, Ginger Grant 2021
A. Moniz et al., *Beginning Azure Cognitive Services*, https://doi.org/10.1007/978-1-4842-7176-6_8

do a better job of recommending movies based on a customer's viewing history. It was a wildly successful contest because it generated a lot of free publicity for offering the contest, and some of the code developed was used to create Apache Spark. Netflix wanted to run the contest again, and when it did, the US Federal Trade Commission got involved. Why? Because two researchers at the University of Texas showed that the data, which was supposed to be anonymized, could be deanonymized relatively easily. A group of Netflix users filed a class action lawsuit for invasion of privacy. The contest was stopped, and Netflix settled the lawsuit for nine million dollars. The lawsuit argued that the data should have been protected under Netflix's privacy policy.

Ethical AI Issues

Data used for analysis is subject to a lot of ethical challenges. As it is easier than ever before to deanonymized data, there have been a number of issues regarding datasets that on the surface should be available to use. Lawsuits have been won even if the data was somewhat publicly available, like the data gathered from the OkCupid website in 2015. That data was extracted from the website using a screen scraping tool by some Danish researchers. They were really surprised at the outcry regarding their use of the data because the data was available to anyone with an OkCupid ID. Once again it was determined that it was not hard to deanonymize the data, making it possible to identify the name of the person who supplied the data. OkCupid complained that the use of the data violated their copyright, and the data was removed from the Internet, after thousands of researchers had done their own analysis of the data.

People being evaluated may notice bias, as was the case in November 2019 with the Goldman Sachs' Apple credit card. Goldman Sachs was accused of bias as high-profile male customers such as Steven Wozniak received a higher credit limit than their female spouses. For customers who filed joint tax returns and had been married for many years, this did not seem correct. Apple blamed the algorithm used to grant credit. Regulators announced they would be investigating Goldman Sachs. Initially Goldman Sachs denied they knew the gender during the application process, a statement that was met by incredulity by regulators and social media accounts. Regulators demanded to know the criteria which were used. Goldman Sachs explained by saying they used personal credit scores and personal debt. When asked to explain the algorithm, Goldman said they could not explain the decision. Senator Elizabeth Warren called for the card to be canceled if the algorithm could not be explained. Such bias issues are not limited to the United States.

These are but three of the more famous examples of issues with data. Issues like these have sparked legislative changes to the use and storage of data, including the General Data Protection Regulation (GDPR) in Europe, which dictates how data can be stored and used. More about that later in this chapter.

Other issues have been found with datasets used for machine learning that adversely impact the results. Loan underwriting has been fraught with problems based upon its datasets. It is illegal to include race or gender as part of a loan algorithm to evaluate credit or cost of credit, but how can regulators ensure that the law is being followed? To get around such laws, proxy discrimination is used. Proxy discrimination is the use of one element as substitute for race or gender. For example, if an organization wishes to include gender as a discriminating factor, instead of asking for the gender, it can be inferred by perhaps reviewing deodorant choice combined with books purchased on Amazon to determine gender as a proxy for actually asking a customers to identify their gender. If using proxy discrimination, the AI might evaluate the type of TV shows a person views. Statistically speaking, you can determine by the types of channels that are selected, for example, Lifetime and BET, which are used as substitutes for race or gender. It is doubtful that such a substitution would be seen as race or gender neutral when evaluated by the legal system.

Governmental Regulation and AI Analysis

It is important to familiarize yourself with the legal issues regarding AI analysis for different industries. For example, while it is illegal to use gender to determine the criteria for underwriting a loan, evaluating gender is allowed in setting rates for insurance in most states within the United States. Gender is used as a large consideration for the rates charged to new drivers, and there is no legal reason against using it. Some US states disagree with this practice and have changed their laws. California has recently made it illegal to allow gender as a criteria to determine insurance rates, which means the algorithms that are used for insurance pricing have changed.

Legal liability with AI is also an issue. In some cases, the software developer is liable; in others, the institutions employing them. Under GDPR, if a machine learning algorithm is used, a person has a right to an explanation. This does not mean that the author of a model that uses Keras or other deep learning tools will have to explain how the algorithm works, but rather what factors contributed to the decision. People who create models will need to provide an explanation of the features and their weight in the decision-making process. As a process perspective, people also need the right to request human intervention, meaning that an AI process will need to be able to be reviewed by a human.

GDPR also mandates the data must be processed fairly so that the data controller can take measures to prevent discriminatory effects on individuals. Data bias needs to be removed in order to comply with the law. If your data is going to be used in a country covered by GDPR, you must be able to provide a human-readable explanation for how the algorithm is used. For this reason, other types of AI may be employed, such as Generative Adversarial Networks (GANs) or federated learning.

A GAN reduces the need for training data. Its main idea is generating input data with the help of output data. Basically, with this method, we take the input and try to figure out what the output will look like. To achieve this, we need to train two neural networks. One is the generator, and the other is the discriminator.

The generator learns how to put data together to generate an image that resembles the output. The discriminator learns how to tell the difference between real data and the data produced by the generator. The problem here is that GANs still require lots of data to be trained properly. This method does not eliminate the need for training data; it just allows us to reduce the amount of initial data and generate a lot of similar augmented data. But if we use a small number of initial datasets, we risk getting a biased AI model in the end. So generative adversarial neural networks do not solve these issues fully, though they do allow us to decrease the quantity of initial testing datasets.

Federated learning is another method of reducing the need for data in AI development. Remarkably, it does not require collecting data at all. In federated learning, personal data does not leave the system that stores it. It's never collected or uploaded to an AI's computers. With federated learning, an AI model trains locally on each system with local data. Later, the trained model merges with the master model as an update. But the problem is that a locally trained AI model is limited, since it's personalized. And even if no data leaves the device, the model is still largely based on personal data. Unfortunately, this contradicts the GDPR's transparency principle, which means federated learning that follows the letter of the law may not be a good choice.

Bias in AI

Bias, which Merriam-Webster defines as an "an inclination of temperament or outlook; especially: a personal and sometimes unreasoned judgment" can be found in the creation of AI algorithms in a number of different ways. One of the more common methods is selection bias. The selection of data elements used in an algorithm or a training decision made by the algorithm based on the data may include elements that are used to reinforce history or stereotypes. Evaluating what criteria one selects when creating an experiment can determine what bias will be learned by the algorithm. Another way bias can creep into an algorithm is through observer bias. If you are expecting a specific result, you may change the selection of data, algorithms, or hyper-parameters to achieve that result. Another method of bias could be survivorship bias. Data that suffers from survivorship bias is data that is previously filtered and then missing some data elements.

One example of a statistical bias was a study about the survivability of cats falling out of buildings. In 1987, veterinarians in New York City's Animal Medical Center did a study on cats that had fallen out of tall buildings. This study concluded that cats that fell 7–32 stories were less likely to die than those that fell 2–6 stories. This study is widely referenced all over the Internet, and many conclusions are drawn from it. But before you go throwing a cat off a tall building, consider this: the data used in this study suffered from statistical bias. In 1997, the data was examined more thoroughly in a syndicated newspaper column called The Straight Dope.

They pointed out if a cat fell off an eight-story building and died, it was not taken to the vet. This meant dead animals were not a part of the study and very few cats that suffered from high-story falls actually were part of this study. Many more cats that fell from shorter distances were taken to the vet and were included in this study, which is why they showed a higher mortality rate. Care must be taken to ensure the data is not filtered in a way that no longer represents the entire group.

Fairness vs. Accuracy

The reason that bias is introduced into the selection of algorithms is many times adding that criteria can improve accuracy. While this is happening, racial or gender bias may also be introduced. By using proxy discrimination, loan underwriters have been caught making it harder for African-Americans to get a low-cost loan. Given the large amount of data that is available through social media accounts, web searches, and credit card purchases, this provides institutions with the ability to evaluate many different data elements to create a digital composite profile. This profile can be used to select a certain kind of borrower while excluding others based upon criteria that treat similarly situated people differently. This may improve accuracy at the expense of an ethical choice. If someone comes from a group that historically has a high default rate, the person may be determined to be high risk, even though that person's personal information would not indicate that decision.

Skewed Samples Compound Issues

There are times that the prediction and decisions made on a given algorithm can influence the environment. Decisions are made that confirm the prediction, and fewer examples are given that contradict the assumption. An example where the analysis has been shown to be a self-fulfilling prophecy is policing by region. If a geographic analysis is done for the locations where there are more people with police records and a machine learning algorithm determines there is more crime in that area, what happens when the local police use that information to send more police to the area? More police means there is a greater chance of arrest, and therefore the algorithm will be seen as correct. If there is an area with the same number of people with police records and police are not sent to the area, will the amount of crime in the area go up? No. There is just a lower probability of contact with the police. How much did the decision made on the data influence the interpretation of the results? This is another area that needs to be considered when selecting data.

Analyzing the Data for Ethical Issues

There are many different issues to evaluate when selecting data. Inferential data such as zip code, social media profiles, or web browsing histories can create groups that reflect historical inequalities. Algorithms need to be analyzed to ensure incorrect conclusions are not made. Data is selected from big datasets, which can be excluded based upon online activities or behaviors. Inferences are made on income based upon these items in combination with zip code, which may not be correct. This can lead to discriminating treatment of individuals or groups, based on race, gender, age, ability, and religion. A Harvard researcher found that this information was used for micro-targeting ads for Black fraternities for high-interest rate credit cards and purchasing arrest records, ignoring the fact that this group has similar background to white groups.

Historical preferences can also result in ethical decisions with current data. In 2015, Amazon created a recruitment tool based on the data gathered from its employees over a 10-year period. The algorithm was supposed to rank resumes to determine the best fit. What they determined when evaluating the tool is it was systematically eliminating women, because the data fed to the algorithm in training included very few women. Resumes that contained things like "Women's Golf Team" were systematically eliminated. The algorithm was selecting resumes that perpetuated the gender mix of the company rather than evaluating skills. The project was abandoned in 2018 when Amazon determined there was no way to eliminate all bias in the algorithm's decision making. AI use in resume selection is still continuing as Goldman Sachs and LinkedIn are still employing AI in the candidate selection process, but state AI is not a replacement for human interaction in the hiring and resume selection process.

Evaluating Training Data

The first factor you will need to evaluate when looking at your training data is data adequacy. Do you have enough data to evaluate the task for the specified purpose? Does your data contain enough different patterns to correctly represent the larger community where the machine learning model will be applied? This may be difficult to evaluate because certain segments of a population may be easier to gather data and as a result the mix of data does not reflect the demographics of the entire group. If you are evaluating people, you will want to review census or similar data for the area in question to make sure you have representative samples of various groups.

There are four factors that are commonly used to evaluate the bias of training data: demographic parity, disparate impact, equal opportunity, and equalized odds. Each of these factors is used to assess bias.

Omitted Variable Bias

Sometimes not including a variable can lead to biased results, most often when doing regression analysis. Sometimes the variable was left off because it was difficult to gather or simply not considered.

A 1999 study in the journal *Nature* determined that leaving a light on in the room of a child 2 years old or younger made it more likely the child would be nearsighted. Researchers at the Ohio State University duplicated the study and found no link between lights and nearsightedness (myopia). They did determine that parents who were more likely to leave the light on in a child's room were those who had nearsightedness, and there was a strong hereditary link between nearsightedness in parents and children. The first study didn't look at the parents' nearsightedness and incorrectly correlated lights as the cause rather than genetics. This study on vision illustrates the bias that can be introduced by leaving out an important data element. Omitted variable bias can cause a correlation without causation.

Correlation and Causation

The mind often looks to find patterns and finds them whether or not the patterns exist outside of the sample set in front of us as that could be an anecdotal occurrence. The pattern could be more complicated than originally it appears. For example, A could cause B only if C happens. Or perhaps A and B are caused by D. Or the causality could be something more complex, such as A causes B only in the case of a long-chain reaction of other events, and you only witnessed the last part. To examine causality, you want to run experiments that control the other variables and then measure the difference. For example, you could test to see if the opposite value is true. For example, let's say you are looking at customer churn and notice a correlation between location and customer turn. If you are able to prove that the location makes a difference to whether or not a customer leaves, you should also be able to prove that a location makes a customer stay. Proving both sides can determine if the result is causation or correlation.

Cognitive Bias

People use cognitive bias to make quick decisions, for example, eating food in gas stations may be bad for your health. This bias may have been established by witnessing hot dogs in many convenience stores sit on rollers forever. Perhaps you remember reading an article about a man who died after eating convenience store nachos in the West United States somewhere. If you live in the northeast of the United States, you may have a different opinion as there is a chain of convenience stores there which is very good. There are even gas stations whose food has been featured on cooking shows. If all of the data is evaluated, perhaps the original bias will be shown to be misplaced as the sample set is expanded. Perhaps you remember the article about the guy dying after eating nachos to help bolster your belief about the quality of their food. If that is the case, you may be suffering from confirmation bias.

There are several different kinds of cognitive bias. In addition to confirmation bias, there is hindsight bias or fundamental attribution error as well. Hindsight bias leads us to believe that we could have predicted the outcome of the past from the future. This makes it difficult to properly evaluate a decision as the outcome is what is focused on, not the process that it takes to arrive at that decision. Confirmation bias occurs when people seek out information that proves the outcome they were anticipating and other results that do not provide that outcome are discarded. For example, an insurance company is trying to figure out if policy holders who have a lot of insurance riders are less sensitive to price increases as they are less likely to leave. The insurance director is pretty sure that this hypothesis is correct and lets the data scientist know that this outcome is practically guaranteed. Initial models that contain data and algorithms that indicate that this is a weak correlation are then discarded and tweaked so the relationship is more prevalent.

Tools for Avoiding Data Bias

When bias is alleged in any AI model, the first step is to interpret the fairness of the algorithm. How does one determine how fair the AI system is? Was there bias in the data used to test and train an algorithm? Do the predictions vary widely when applied to different groups?

Fairlearn

Microsoft created the open source toolkit Fairlearn in 2019 to evaluate how fair an algorithm is and to mitigate the factors identified in the algorithm. The analysis is performed in a Python package that can be used to assess a system's fairness and mitigate fairness issues. Fairlearn determines fairness by evaluating allocation harms and quality of service harms. Allocation harms are defined as situations where the AI systems withhold opportunities or resources in areas such as hiring, lending or school admissions. Quality of service harms are meant to measure if the system works well for different people even if no opportunities result as a part of the decision-making process. For example, voice recognition works for female voices but does not work for males. This requires that the principle of group fairness be applied, which expects that the data scientist will identify groups of users who can be represented in the data.

Fairlearn Dashboards

Fairlearn contains a dashboard with different elements used to determine tradeoffs between performance and fairness. The visualizations can help identify the impact of the model on groups of data using a number of different metrics. The dashboard is a Jupyter Notebook widget designed to evaluate how predictions impact different groups. Here is the Python code to create the dashboard using version v.0.5.0:

```python
from fairlearn.widget import FairlearnDashboard

FairlearnDashboard(sensitive_features=A_test,
                   sensitive_feature_names=['Race', 'Age'],
                   y_true=Y_test.tolist(),
                   y_pred=[y_pred.tolist()])
```

The variable A_test is used to evaluate the sensitive values of your data, and the sensitive features are listed in the names. Y_true contains the ground truth labels, and y_pred contains the label predictions. The dashboard walks the user through the assessment setup, where the sensitive features are to be evaluated, and the performance metric to evaluate the model performance as well as any disparities between groups. For example, if your data contained elements of race and gender, you would be prompted to select the different groups to evaluate. Next, you will be prompted for how you want to measure performance. After this setup, you will see the following screen to evaluate the disparity in the algorithm performance. Figure 8-1 shows the dashboard created by Fairlearn to show potential issues with the data.

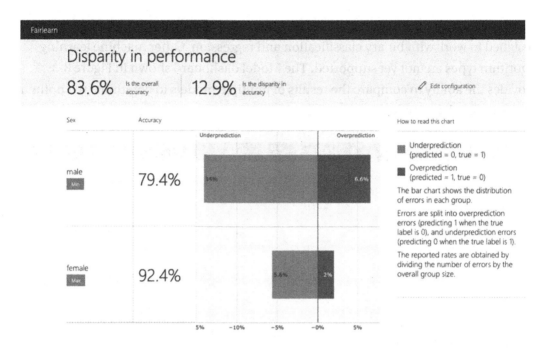

Figure 8-1. *Fairlearn performance disparity dashboard*

As shown in Figure 8-2, there is also an additional dashboard to display the disparity in prediction results between the different identified groups.

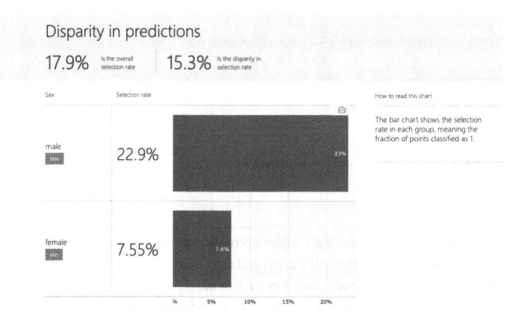

Figure 8-2. *Fairlearn disparity in predictions dashboard*

There are two types of mitigation algorithms used to mitigate fairness, which are designed to work with binary classification and regression. Other machine learning algorithm types are not yet supported. The Model dashboard shown in Figure 8-3 provides the ability to compare the results of multiple models to determine the optimal choice for model selection.

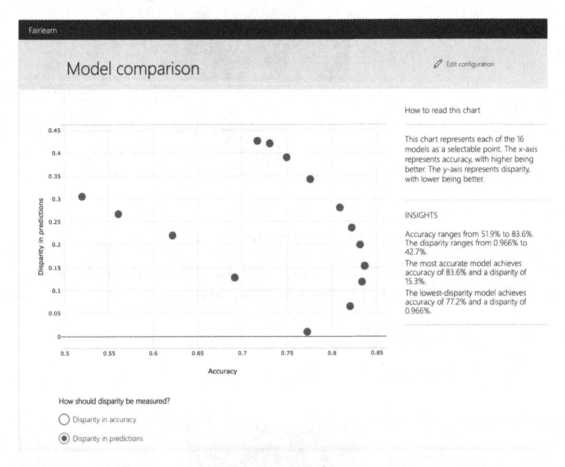

Figure 8-3. *Fairlearn model comparison visual*

Mitigation

Once a model has been evaluated, Fairlearn provides some different methods for resolving disparity for binary classification and regression algorithms. Wrappers are provided to implement the reduction approach to fair classification and equality of opportunity in supervised learning. The mitigation algorithms are models after Scikit-Learn as Fairlearn implements the fit model to train a model and the predict model to make predictions. There are three algorithms used in mitigating unfairness, Exponentiated Gradient, GridSearch, and ThresholdOptimizer. These algorithms use a wrapper approach to fit the provided classifiers lending to reduced differences in the selection rate.

InterpretML

Another important solution is InterpretML, a tool and set of interactive dashboards that uses various techniques to deliver model explanations at inference. For different types of models, InterpretML helps practitioners better understand the most important features in determining the model's output, perform "what if" analyses, and explore trends in the data. Microsoft also announced it was adding a new user interface equipped with a set of visualizations for interpretability, support for text-based classification, and counterfactual example analysis to the toolset.

InterpretML is an open source Python toolkit for training interpretable glass box models and explaining black box systems. InterpretML helps you understand your model's global behavior or understand the reasons behind individual predictions. It includes a built-in dashboard to provide visual information on datasets, model performance, and model explanations.

The InterpretML toolkit can help provide answers regarding machine learning including

- *Model debugging* – Why did my model make this mistake?

- *Detecting fairness issues* – Does my model discriminate?

- *Human-AI cooperation* – How can I understand and trust the model's decisions?

- *Regulatory compliance* – Does my model satisfy legal requirements?

InterpretML uses a Explainable Boosting Machine (EBM), an interpretable model developed at Microsoft Research, to evaluate models. It uses modern machine learning techniques like bagging, gradient boosting, and automatic interaction detection to breathe new life into traditional GAMs (Generalized Additive Models). This makes EBMs as accurate as state-of-the-art techniques like random forests and gradient-boosted trees. However, unlike these black box models, EBMs produce lossless explanations and are editable by domain experts.

InterpretML provides a dashboard to help the user understand model performance for different subsets of data, explore model errors, and access dataset statistics and distributions. It includes summary explanations as well as overall and individual explanations, using a variety of different methods and enables performing feature perturbations via what if analysis exploring how model predictions change as features are perturbed. Figure 8-4 shows an example of the dashboard created to analyze the model performance.

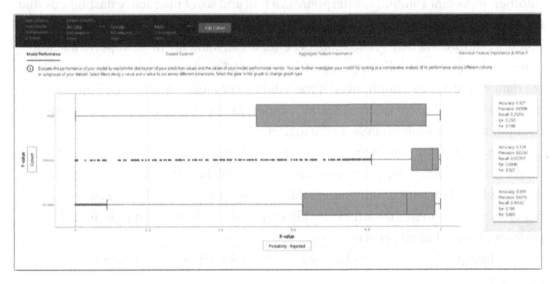

Figure 8-4. *InterpretML dashboard*

InterpretML is further extended by an experimental repository method called Interpret-Community.

Interpret-Community

Actively incorporates innovative experimental interpretability techniques and allows for further expansion by researchers and data scientists.

Applies optimizations to make it possible to run interpretability techniques on real-world datasets at scale.

Provides improvements such as the capability to "reverse the feature engineering pipeline" to provide model insights in terms of the original raw features rather than engineered features.

Provides interactive and exploratory visualizations to empower data scientists to gain meaningful insight into their data.

Differential Privacy

Differential privacy is a class of algorithms that facilitate computing and statistical analysis of sensitive, personal data while ensuring the privacy of individuals isn't compromised. Microsoft unveiled WhiteNoise, a library of open source algorithms that enable machine learning on private, sensitive data. Differential privacy is a system for publicly sharing information about a dataset by describing the patterns of groups within the dataset while withholding information about individuals in the dataset. The idea behind differential privacy is that if the effect of making an arbitrary single substitution in the database is small enough, the query result cannot be used to infer much about any single individual and therefore provides privacy. Another way to describe differential privacy is as a constraint on the algorithms used to publish aggregate information about a statistical database, which limits the disclosure of private records whose information is in the database. For example, differentially private algorithms are used by some government agencies to publish demographic information or other statistical aggregates while ensuring confidentiality of survey responses and by companies to collect information about user behavior while controlling what is visible even to internal analysts.

Differential privacy is a formal mathematical framework for guaranteeing privacy protection when analyzing or releasing statistical data. Recently emerging from theoretical computer science literature, differential privacy is now in initial stages of implementation and use in various academic, industry, and government settings. Using intuitive illustrations and limited mathematical formalism, this document provides an introduction to differential privacy for nontechnical practitioners, who are increasingly tasked with making decisions with respect to differential privacy as it grows more widespread in use. In particular, the examples in this document illustrate ways in which social scientists can conceptualize the guarantees provided by differential privacy with respect to the decisions they make when managing personal data about research subjects, informing them about the privacy protection they will be afforded.

Differential privacy uses two steps to achieve privacy benefits. First, a small amount of statistical noise is added to each result to mask the contribution of individual data points. The noise is significant enough to protect the privacy of an individual, but still small enough that it will not materially impact the accuracy of the answers extracted by analysts and researchers. Next, the amount of information revealed from each query is calculated and deducted from an overall privacy loss budget, which will stop additional queries when personal privacy may be compromised. This can be thought of as a built-in shutoff switch in place that prevents the system from showing data when it may begin compromising someone's privacy.

Deploying Differential Privacy

Differential privacy has been a vibrant area of innovation in the academic community since the original publication. However, it was notoriously difficult to apply successfully in real-world applications. In the past few years, the differential privacy community has been able to push that boundary, and now there are many examples of differential privacy in production.

Differential Privacy Platform

This project aims to connect theoretical solutions from the research community with the practical lessons learned from real-world deployments, to make differential privacy broadly accessible. The system adds noise to mask the contribution of any individual data subject and thereby provide privacy. Figure 8-5 shows a diagram on the different elements of differential privacy.

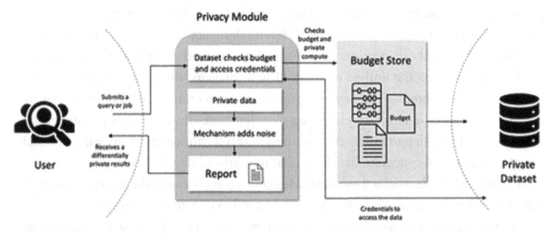

Figure 8-5. *Differential privacy*

It is designed as a collection of components that can be flexibly configured to enable developers to use the right combination for their environments. The core includes a pluggable open source library of differentially private algorithms and mechanisms with implementations based on mature differential privacy research. The library provides a fast, memory-safe native runtime. In addition, there are APIs for defining an analysis and a validator for evaluating this analysis and composing the total privacy loss on a dataset. The runtime and validator are built in Rust, while Python support is available.

Summary

In this chapter, we reviewed several different issues with data analysis and looked beyond the technical elements of AI to explore how the results are used. We reviewed some ethical issues that were raised when including data where the users did not consent to be analyzed. We looked at results from studies where incomplete datasets led to erroneous conclusions. After examining some of the problems, we looked at technology that could be employed to solve some of the issues with data used in machine learning by examining how to use Fairlearn and InterpretML and how they could be used to improve analysis.

Index

A

AI Builder
    data objects
        detection, 238
        PowerApps, 239
        tagging, 239
        uses, 239
    prebuilt AI models
        Auto map, 233
        CategorizeText function, 225
        CDS, 230
        COVID-19, 223
        create solution, 225, 226
        CSV file, 232, 233
        customization, 229
        Designer, 237
        explanation, 230
        Get data, 232
        importing, 232
        onlineshopperIntention, 231
        outcomes, 235
        PowerApps, 224, 234
        Power Automate
            flow, 226, 227
        predictions, 231
        published view, 237, 238
        responses, 228
        results, 236
        sentiment analysis, 223, 224
        SharePoint list, 227, 228

        solution package, 231
        training, 235, 236
        Views, 234
        Vision category, 229
    scenarios, 223
AI engineer toolkit
    Azure AI gallery, 25
    Azure data studio
        download, 21
        interface, 21, 22
        vs. SSMS, 22, 23
    machine/deep learning
        frameworks
        CNTK, 25
        Pandas, 23
        Scikit-Learn, 24
        Spark MLlib, 24
        TensorFlow, 24, 25
AI ethics
    bias, 284, 285
    cognitive bias, 288, 289
    correlation/causation, 288
    data analyzing, 286, 287
    evaluating data, 287
    fairness vs. accuracy, 285
    government regulation, 283, 284
    issues, 281–283
    omitted variable bias, 287
    skewed samples compound
        issues, 286

Printed in the United States
by Baker & Taylor Publisher Services